D0015860

ALSO BY ALAN BURDICK

Out of Eden

WHY TIME FLIES

A MOSTLY SCIENTIFIC INVESTIGATION

ALAN BURDICK

Simon & Schuster

NEW YORK   LONDON   TORONTO   SYDNEY   NEW DELHI

Simon & Schuster
1230 Avenue of the Americas
New York, NY 10020

First Simon & Schuster hardcover edition February 2017

SIMON & SCHUSTER and colophon are
registered trademarks of Simon & Schuster, Inc.

For information about special discounts for bulk purchases,
please contact Simon & Schuster Special Sales at
1-866-506-1949 or business@simonandschuster.com.

The Simon & Schuster Speakers Bureau can bring authors to your
live event. For more information or to book an event contact the
Simon & Schuster Speakers Bureau at 1-866-248-3049
or visit our website at www.simonspeakers.com.

Illustration Credits
All illustrations by Stephen Burdick Design with the exception of the following:
Courtesy of Ernst Pöppel, from *Mindworks:*
Time and Conscious Experience: 142, 143, 144
Courtesy of Museu de Arte de São Paulo Assis Chateaubriand: 210
Courtesy of the Trustees of the Boston Public Library/Rare Books: 151, 152

Manufactured in the United States of America

1 3 5 7 9 10 8 6 4 2

Library of Congress Cataloging-in-Publication Data

Names: Burdick, Alan.
Title: Why time flies : a mostly scientific investigation / by Alan Burdick.
Description: New York : Simon & Schuster, [2017] | Includes bibliographical
references and index.
Identifiers: LCCN 2016025791 (print) | LCCN 2016026733 (ebook) |
ISBN 9781416540274 (hardcover : alk. paper) |
ISBN 9781416540281 (pbk. : alk. paper) | ISBN 9781451677010 (ebook)
Subjects: LCSH: Time measurements—Popular works.
Classification: LCC QB213 .B925 2017 (print) | LCC QB213 (ebook) |
DDC 529/.2—dc23
LC record available at https://lccn.loc.gov/2016025791

ISBN 978-1-4165-4027-4
ISBN 978-1-4516-7701-0 (ebook)

for Susan

I confess to you, Lord, that even today I am still ignorant of what time is; but I praise you, Lord, for the fact that I know I am making this avowal within time, and for my realization that within time I am talking about time at such length, and that I know this "length" itself is long only because time has been passing all the while.

—St. Augustine, *The Confessions*

One of the girls devised a method of stamping envelopes which enabled her to work at a speed of between one hundred and one hundred and twenty envelopes per minute. . . . We do not know just what processes were followed in developing the method, as the girl studied it out and put it in operation while the writer was taking a vacation.

—Frank Gilbreth, *Motion Study:*
A Method for Increasing the Efficiency of the Workman

CONTENTS

FORWARD

Some nights—more than I like, lately—I wake to the sound of the bed-side clock. The room is dark, without detail, and in darkness the room expands in such a way that it seems as if I am outdoors, under an endless empty sky, yet at the same time underground, in a vast cavern. I might be falling through space. I might be dreaming. I could be dead. Only the clock moves, its tick steady, unhurried, relentless. At these moments I have the clearest and most chilling understanding that time moves in one direction only.

In the beginning, or right before that, there was no time. According to cosmologists, the universe started nearly fourteen billion years ago with a "big bang" and in an instant expanded to something closer to its current size, and it continues to expand faster than the speed of light. Before all that, though, there was nothing: no mass, no matter, no energy, no gravity, no motion, no change. No time.

Maybe you can imagine what that was like. I can't fathom it. My mind refuses to receive the idea and instead insists, Where did the universe come from? How does something appear from nothing? For argument's sake, I'll accept that perhaps the universe did not exist before the Big Bang—but it exploded *in* something, right? What was that? What was there before the beginning?

Proposing such questions, the astrophysicist Stephen Hawking has said, is like standing at the South Pole and asking which way is south: "Earlier times simply would not be defined." Perhaps Hawking is trying to be reassuring. What he seems to mean is that human language has a limit. We (or at least the rest of us) reach this boundary whenever we ponder the cosmic. We imagine by analogy and metaphor: that strange and vast thing is like this smaller, more familiar thing. The universe is a cathedral, a clockworks, an egg. But the parallels ultimately diverge; only an egg is an egg. Such analogies appeal precisely because they are tangible elements of the universe. As terms, they are self-contained—but they cannot contain the container that holds them.

So it is with time. Whenever we talk about it, we do so in terms of something lesser. We find or lose time, like a set of keys; we save and spend it, like money. Time creeps, crawls, flies, flees, flows, and stands still; it is abundant or scarce; it weighs on us with palpable heft. Bells toll for a "long" or a "short" time, as if their sound could be measured with a ruler. Childhood recedes, deadlines loom. The contemporary philosophers George Lakoff and Mark Johnson have proposed a thought experiment: take a moment and try to address time strictly in its own terms, stripped of any metaphor. You'll be left empty-handed. "Would time still be time for us if we could not *waste* or *budget* it?" they wonder. "We think not."

Begin with a word, as God did, Augustine urges the reader: "You spoke and things were made. By your word you made them."

The year is 397. Augustine is forty-three, midway through his life as an overwhelmed bishop in Hippo, a North African port city of the fallen Roman Empire. He has written dozens of books—collections of sermons, scholarly rebukes of his theological foes—and now undertakes *The Confessions*, a strange and riveting work that will take four years to complete. In the first nine of its thirteen chapters, Augustine recounts the key details of his life from infancy (as best he can infer) to his formal embrace of Christianity, in 386, and his mother's death the following year. Along the way he accounts for his sins, among them thievery (he stole pears from a neighbor's tree), sex out of wedlock, astrology, fortune-telling, superstitions, an interest in theater, and more sex. (Actually, Augustine was monogamous for most of his life, first to a longtime companion and later to a wife by arranged marriage, after which he turned chaste.)

The remaining four chapters are something else entirely: an extended meditation on, in ascending order, memory, time, eternity, and Creation. Augustine is frank about his ignorance of the divine and natural order and dogged in his pursuit of clarity. His conclusions and his introspective method would inform centuries of subsequent philosophers, from Descartes (whose *cogito ergo sum*—I think therefore I am—is a direct echo of Augustine's *dubito ergo sum*, I doubt therefore I am) to Heidegger to Wittgenstein. He grapples with the Beginning: "I will set about replying to the questioner who asks, 'What was God doing before he made heaven and earth?' But I will not respond with that joke someone is said to have made: 'He is getting hell ready for people who inquisitively peer into deep matters.'"

Augustine's *Confessions* is sometimes described as the first true autobiography—a self-told story of how a self grew and changed with time. I have come to think of it as a memoir of evasion. In the early chapters divinity knocks but Augustine won't answer. He fathers an illegitimate son; while studying rhetoric in Rome he takes up with a group of rabble-rousing friends he calls "the wreckers"; his devout mother frets over his wayward lifestyle. Augustine later describes this period of his life as "no more than anxious distraction." His *Confessions* manifests what we have come to embrace as a thoroughly modern idea, familiar to anyone acquainted with psychotherapy: that one's scattered past can be reformed into a meaningful present. Your memories are yours and through them you can shape for yourself a new narrative that enlightens and defines you. "That from days of prior dispersion I may collect myself into identity," Augustine writes. It is autobiography as self-help. *Confessions* is a book about many things, chief among them words and their capacity, through time, to redeem.

For a long time, time was something I did my best to avoid. For instance, for much of my early adulthood I refused to wear a watch. I'm not quite sure how I landed on that decision; I vaguely recall reading that Yoko Ono never wears one because she hates the notion of having time strapped to her wrist. That made sense. Time, it seemed to me, was an external phenomenon, imposed and oppressive—and therefore something I could actively choose to remove from my person and leave behind.

This notion initially gave me a deep sense of pleasure and relief, as rebellions often do. It also usually meant that, as I headed off somewhere or to meet someone, I was not outside time at all, I was simply behind it. I was late. I was so effective at avoiding time that a long time passed before I understood that that's what I was doing. And with that realization, another one quickly followed: I was avoiding time because secretly I feared it. I gained a sense of control from perceiving time as external, as if it were something I could step in and out of, like a stream, or sidestep altogether, like a lamppost. But deep down I sensed the truth: time was—is—*in* me, in us. It is there from the moment I wake to the moment I fall asleep, it suffuses the air, it permeates the mind and body, it crawls through one's cells, through every living moment, and will continue advancing long after the moment it leaves all cells behind. I felt infected. And yet I could not say where it came from, much less where it went—and keeps going, steadily leaking away. As with so many things that one fears, I had no idea what time actually is, and my skill at avoiding it only led me further from any real answer.

And so one day, longer ago than I wish were true, I set out on a journey through the world of time in order to understand it—to ask, as Augustine did, "Where is it coming from, what is it passing through, and where is it going?" The purer physical and mathematical aspects of time continue to be debated by the great minds of cosmology. What interested me, and what science has only begun to reveal, is how time manifests itself in living biology: how it is interpreted and told by cells and subcellular machinery, and how that telling seeps upward into the neurobiology, psychology, and consciousness of our species. As I traveled through the world of time research and visited with its many ologists, I sought answers to questions that have long plagued me and perhaps you too, such as, Why did time seem to last longer when we were children? Does the experience of time really slow down when you're in a car crash? How is it that I'm more productive when I have too much to do, whereas when I have all the time in the world, I seem to get nothing done? Is there a clock in us that counts off the seconds, hours, and days, like the clock in a computer? And if we contain such a clock, how pliable is it? Can I make time speed up, slow down, stop, reverse? How and why does time fly?

I can't say exactly what I was after—peace of mind, maybe, or some insight into what Susan, my wife, once referred to as my "willful denial of the passage of time." For Augustine, time was a window onto the soul. Modern science is more concerned with probing the framework and texture of consciousness, a concept that is only slightly less elusive. (William James dismissed consciousness as "the name of a nonentity . . . a mere echo, the faint rumor left behind by the disappearing 'soul' on the air of philosophy.") Yet whatever one calls it, we share a rough idea of what's meant: a lasting sense of one's self moving in a sea of selves, dependent yet alone; a sense, or perhaps a deep and common wish, that *I* somehow belongs to *we*, and that this *we* belongs to something even larger and less comprehensible; and the recurring thought, so easy to brush aside in the daily effort to cross the street safely and get through one's to-do list, much less to confront the world's true crises, that my time, our time, matters precisely because it ends.

I imagined a meditation, then, with luck a reckoning. I should mention here that my previous book took me far longer to write than I had intended or even imagined possible. So I made a vow to myself: I would undertake a new book only on the condition that I would absolutely finish it on time—and in a reasonable amount of time at that. In effect, *Why Time Flies* would be a book about time, written on time. Of course, it wasn't. What started as a journey evolved into something between a pastime and an obsession, accompanying me through one job and another, the birth of my children, preschool, grade school, beach vacations, and canceled deadlines and dinner dates; in its sway I beheld the most accurate clock in the world, experienced the white nights of the Arctic, and fell from a great height into the arms of gravity. My subject settled in for the long haul, a hungry houseguest, beguiling and instructive, much like time itself.

I'd scarcely begun when I met a fundamental fact about time: there is no one truth about it. Instead, I found a multitude of scientists across the spectrum of time research; each could speak confidently about his or her narrow wavelength but none could say quite how it all adds up to white light or what that looks like. "Just when you think you understand what's going on," one told me, "there's another experiment that changes one small aspect, and suddenly you don't know what's going on again." If scientists agree on anything, it's that nobody knows enough about time and that this lack of knowledge is surprising given how pervasive and integral time is to our lives. Another researcher confided, "I can imagine one day aliens coming down from outer space and saying, 'Oh, actually, time is such and such,' and we'll all nod our heads like it was obvious the whole time." If anything, time seemed to me to be a lot like the weather: something that everyone talks about but never does anything about. I intended to do both.

THE HOURS

One can expect an agreement between philosophers sooner than between clocks.

—Seneca, *The Pumpkinification of Claudius*

I settle into a seat on the Paris Métro and rub the sleep from my eyes. I feel unmoored. The calendar says late winter but outside my window the day is warm and fair, the leaf buds gleam, the city is resplendent. I arrived from New York yesterday and stayed out past midnight with friends; today my head is still in the dark, glued in a season and a time zone several hours behind me. I glance at my watch: 9:44 a.m. As usual, I am late.

The watch is a recent gift from my father-in-law, Jerry, who wore it himself for many years. When Susan and I became engaged, her parents offered to buy me a new watch. I declined, but for a long time afterward I couldn't shake the worry that I'd made a poor impression. What sort of son-in-law ignores the time? So when Jerry subsequently offered me his old wristwatch I said yes right away. It has a golden dial set on a wide silver wristband; a black face bearing the brand name (Concord) and the word *quartz* in bold letters; and the hours denoted by unnumbered lines. I liked the new weight on my wrist, which made me feel important. I thanked him and remarked, more accurately than I could understand at that moment, that it would be a helpful addition to my research on time.

On the evidence of my senses, I had come to believe that the time "out there" in clocks, watches, and train schedules is quantifiably distinct from the time coursing through my cells, body, and mind. But the fact was that I knew as little about the former as I did about the latter. I could not say how a particular clock or watch worked nor how it managed to agree so closely with the other watches and clocks that I occasionally noticed. If there was a real difference between external and internal time—as real as the difference between physics and biology—I had no idea what it was.

So my new, used watch would be a kind of experiment. What better way to plumb my relationship to time than to physically attach it to me for a while? Almost immediately I saw results. For the first few hours of wearing the watch I could think about nothing else. It made my wrist sweat and tugged at my whole arm. Time dragged literally and, because my mind dwelt on the dragging, figuratively. Soon enough I forgot about the watch. But on the evening of the second day I suddenly remembered it again when, while bathing one of our infant sons in the tub, I noticed it on my wrist, underwater.

Secretly I hoped that the watch might confer some degree of punctuality. For instance, it seemed to me that if I looked at the watch often enough I might yet arrive on time for my ten o'clock appointment in Sèvres, just outside Paris, at the Bureau International des Poids et Mesures—the International Bureau of Weights and Measures. The Bureau is an organization of scientists devoted to perfecting, calibrating, and standardizing the basic units of measurement used around the world. As our economies globalize, it becomes ever more imperative that we all be on precisely the same metrological page: that one kilogram in Stockholm equals exactly one kilogram in Jakarta, that one meter in Bamako equals exactly one meter in Shanghai, that one second in New York equals exactly one second in Paris. The Bureau is the United Nations of units, the world standardizer of standards.

The organization was formed in 1875 through the Convention of the Metre, a treaty meant to ensure that the basic units of measurement are uniform and equivalent across national borders. (The first act of the Convention was for the Bureau to hand out rulers: thirty precisely measured bars made of platinum and iridium, which would settle international disagreements over the correct length of a meter.) Seventeen nation members joined the original Bureau; fifty-eight now belong, including all the major industrialized nations. The suite of standard units it oversees has grown to seven: the meter (length), the kilogram (mass), the ampere (electrical current), the kelvin (temperature), the mole (volume), the candela (luminosity), and the second.

Among its many duties, the Bureau maintains a single, official worldwide time for all of Earth, called Coordinated Universal Time, or U.T.C. (When U.T.C. was first devised, in 1970, the organizing parties could not agree on whether to use the English acronym, C.U.T., or the French

4

acronym, T.U.C., so they compromised on U.T.C.) Every timepiece in the world, from the hyperaccurate clocks in orbiting global-positioning satellites to the cog-bound wristwatch, is synchronized directly or eventually to U.T.C. Wherever you live or go, whenever you ask what time it is, the answer ultimately is mediated by the timekeepers at the Bureau.

"Time is what everybody agrees the time is," a time researcher explained to me at one point. To be late, then, is to be late according to the agreed-on time. By definition, the Bureau's time is not merely the most correct time in the world, it is precisely the correct time. This meant, as I glanced at my watch yet again, that I was not merely late: I was as late as I have ever been and as late as it is possible to be. Soon enough I would learn just how far behind the time I truly was.

A clock does two things: it ticks and it counts the ticks. The clepsydra, or water clock, ticks to the steady drip of water, which, in more advanced devices, drives a set of gears that nudges a pointer along a series of numbers or hash marks, thereby indicating time's passage. The clepsydra was in use at least three thousand years ago, and Roman senators used them to keep their colleagues from talking for too long. (According to Cicero, to "seek the clock" was to request the floor and to "give the clock" was to yield it.) Water ticked and added up to time.

For most of history, though, in most clocks, what ticked was Earth. As the planet rotates on its axis, the sun crosses the sky and casts a moving shadow; cast on a sundial, the shadow indicates where you are in the day. The pendulum clock, invented in 1656 by Christiaan Huygens, relies on gravity (affected by Earth's rotation) to swing a weight back and forth, which drives a pair of hands around the face of the clock. A tick is simply an oscillation, a steady beat; Earth's turning provided the rhythm.

In practice, what ticked was the day, the rotational interval from one sunrise to the next. Everything in between—the hours and minutes— was contrived, a man-made way to break up the day into manageable units for us to enjoy, employ, and trade. Increasingly our days are governed by seconds. They are the currency of modern life, the pennies of our time: ubiquitous and critical in a pinch (for instance, when you just manage to make a train connection) yet sufficiently marginal to be frittered away or dropped by the handful without thought. For centuries, the second existed only in the abstract. It was a mathematical subdivision, defined by relation: one-sixtieth of a minute, one thirty-six-hundredth of an hour, one eighty-six-thousand-four-hundredth of a day. Seconds pendulums appeared on some German clocks in the fifteenth century. But it wasn't until 1670, when the British clockmaker William Clement added a seconds pendulum, with its familiar tick-tock, to Huygens's pendulum clock, that the second acquired a reliably physical, or at least audible, form.

The second fully arrived in the twentieth century, with the rise of the quartz clock. Scientists had found that a crystal of quartz resonates like a tuning fork, vibrating at tens of thousands of times per second when placed in an oscillating electrical field; the exact frequency depends on the size and shape of the crystal. A 1930 paper titled "The Crystal Clock" noted that this property could drive a clock; its time, derived from an electrical field instead of gravity, would prove reliable in earthquake zones and on moving trains and submarines. Modern quartz clocks and wristwatches typically use a crystal that has been laser-engineered to vibrate at exactly 32,768 (or 2^{15}) times per second, or 32,768 Hz. This provided a handy definition of the second: 32,768 vibrations of a quartz crystal.

By the nineteen-sixties, when scientists managed to measure an atom of cesium naturally undergoing 9,192,631,770 quantum vibrations per second, the second had been officially redefined to several more decimal places of accuracy. The atomic second was born, and time was upended. The old temporal scheme, known as Universal Time, was top-down: the second was counted as a fraction of the day, which took its shape from Earth's motion in the heavens. Now, instead, the day would be measured from the ground up, as an accumulation of seconds. Philosophers debated whether this new atomic time was as "natural" as the old time. But there was a bigger problem: the two times don't quite agree. The increasing accuracy of atomic clocks revealed that Earth's rotation is gradually slowing, adding very slightly to the length of each day. Every couple of years this slight difference adds up to a second; since 1972, nearly half a minute's worth of "leap seconds" have been added to International Atomic Time to bring it into sync with the planet.

In the old days, anyone could make his or her own seconds through simple division. Now the seconds are delivered to us by professionals; the official term is "dissemination," suggesting an activity akin to gardening or the distribution of propaganda. Around the world, mainly in national timekeeping laboratories, some three hundred and twenty cesium clocks, each the size of a small suitcase, and more than a hundred large, maser-driven devices generate, or "realize," highly accurate seconds on a near-continuous basis. (The cesium clocks, in turn, are checked against a frequency standard generated by a device called a cesium fountain—a dozen or so exist—which uses a laser to toss cesium atoms around in a vacuum.) These realizations are then added up to reveal the time of day. As Tom Parker, a former group leader at the National Institute of Standards and Technology, told me, "The second is the thing that ticks; time is the thing that counts the ticks."

N.I.S.T. is a federal agency that helps produce the official, civil time for the United States. Experts at its two laboratories, in Gaithersburg, Maryland, and Boulder, Colorado, keep a dozen or more cesium clocks running at any given time. As precise as these clocks are, they disagree with one another on a scale of nanoseconds, so every twelve minutes they are compared to one another tick by tick to see which are running fast and which are running slow and by exactly how much. The data from the clock ensemble is then numerically mashed into what Parker calls "a fancy average," and this becomes the basis for the official time.

How this time reaches you depends on your timekeeping device and where you happen to be at the moment. The clock in your laptop or computer regularly checks in with other clocks across the Internet and calibrates itself to them; some or all of these clocks eventually pass through a server run by N.I.S.T. or another official clock and are thereby set even more accurately. Every day, N.I.S.T.'s many servers register 13 billion pings from computers around the world inquiring about the correct time. If you are in Tokyo, you might be linked to a time server in Tsukuba that is run by the National Metrology Institute of Japan; in Germany, the source is the Physikalisch-Technische Bundesanstalt.

Wherever you are, if you're checking the clock on your cell phone, it's probably receiving its time from the Global Positioning System, an array of navigation satellites synchronized to the U.S. Naval Observatory, near Washington, D.C., which realizes its seconds with an ensemble of seventy-odd cesium clocks. Many other clocks—wall clocks, desk clocks, wristwatches, travel alarms, car-dashboard clocks—contain a tiny radio receiver that, in the United States, is permanently tuned to pick up a signal from N.I.S.T. Radio Station WWVB, in Fort Collins, Colorado, which broadcasts the correct time as a code. (The signal is very low frequency—60 Hz—and the bandwidth so narrow that a good minute is needed for the complete time code to come through.) These clocks can generate the time on their own, but for the most part they act as middlemen, serving you the time that is disseminated by more refined clocks somewhere higher up in the temporal chain of command.

My wristwatch, in contrast, has no radio receiver or any way of talking to satellites; it's all but off the grid. To synchronize with the wider world I need to look at an accurate clock and then turn the stem of my watch and set the time accordingly. To achieve even greater accuracy I could regularly take my watch to a shop and have its mechanism calibrated to a device called a quartz oscillator, which gains its precision from a frequency standard monitored by N.I.S.T. Otherwise, my watch will keep its realizations to itself and will soon fall out of step with everyone else's. I had assumed that putting on a watch meant strapping established time to my wrist. But, in fact, unless I take the measure of the clocks around me, I am still a rogue. "You're free-running," Parker said.

From the late seventeenth century to the early twentieth century, the most accurate clock in the world resided at the Royal Observatory in Greenwich, England; it was regularly reset by the Astronomer Royal according to the movement of the heavens. This situation was good for the world but quickly became a problem for the Astronomer Royal. Beginning around 1830, he increasingly found himself interrupted from his work by a knock on the door from a townsperson. *Pardon me*, he was asked. *Would you tell me the time?*

So many people came knocking that eventually the town petitioned the astronomer for a proper time service; in 1836 he assigned his assistant, John Henry Belville, to the task. Every Monday morning, Belville calibrated his timepiece, a pocket chronometer originally made for the duke of Sussex by the esteemed clockmaker John Arnold & Son, to the observatory time. Then he set off for London to visit his clients— clockmakers, watch repairers, banks, and private citizens who paid a fee to synchronize their time to his and, by extension, the observatory's. (Belville eventually replaced the chronometer's gold case with a silver one in order to draw less attention in "the less desirable quarters of the town.") When Belville died, in 1856, his widow took over; when she retired, in 1892, the service passed to their daughter Ruth, who became known as "the Greenwich time lady." Using the same chronometer, which she called "Arnold 345," Miss Belville made the same tour, disseminating what by then was known as Greenwich Mean Time, the official time of Britain. The invention of the telegraph, which enabled remote clocks to synchronize with Greenwich time almost immediately and at lower cost, eventually rendered Miss Belville almost but not quite obsolete. When she retired around 1940, in her mideighties, she still served some fifty clients.

I had come to Paris to meet with the Greenwich time lady of the modern era, the Miss Belville for all of Earth: Dr. Elisa Felicitas Arias, the director of the B.I.P.M.'s Time Department. Arias is slender, with long brown hair and the air of a kindly aristocrat. An astronomer by training, Arias worked for twenty-five years at observatories in Argentina, her native country, the last ten of them with the Naval Observatory; her specialty is astrometry, the correct measuring of distances in outer space. Most recently she worked with the International Earth Rotation and Reference Systems Service, which monitors the ever-so-slight variations in our planet's motions and consequently determines when the next leap second should be added to the temporal mix. I met her in her office, and she offered me a cup of coffee. "We have one common objective," she said of her department. "To provide a timescale suitable to be an international reference." The aim, she added, is "ultimate traceability."

Of the hundreds of clocks and clock ensembles run by the Bureau's fifty-eight member-nations, only about fifty—the "master clocks," one per country—are up and running and providing official time; everywhere, at all hours, they realize seconds. But their realizations don't agree with one another. It's a matter of nanoseconds, or billionths of a second. That's not enough to trouble electrical-power companies (which need accuracies only in the milliseconds) or disrupt telecommunications (which traffic in microseconds). But the clocks on different navigation systems—such as G.P.S., which is run by the U.S. Department of Defense, and the European Union's new Galileo network—need to agree within a few nanoseconds in order to provide consistent service. The world's clocks should agree, or should at least be well aimed toward the same point of synchrony, and Universal Coordinated Time is the designated goal.

Universal Coordinated Time is derived by comparing all the member clocks as they tick their seconds simultaneously, and noting the discrepancies. It is a tremendous technical challenge. For one thing, the clocks are hundreds or thousands of miles apart. Given the time it takes for an electronic signal to traverse such distances—a signal that says, in effect, "Start ticking now"—it is difficult to know precisely what "at the same time" means. To get around this problem, Arias's section uses G.P.S. satellites to transfer data. The satellites all have known positions and carry clocks synchronized to the U.S. Naval Observatory; with this information, the B.I.P.M. can calculate the precise moments when time signals are being sent to them from clocks around the world.

Even then, uncertainties loom. The position of a satellite can't be known exactly; bad weather and Earth's atmosphere can slow or alter a signal's path and obscure its true travel time. And the equipment harbors electronic noise that can obscure precise measurement. Offering an analogy, Arias motioned to the door of her office. "If I ask you what time it is, you'll tell me the time and I'll compare it to mine," she said. "We are face-to-face. If I say, 'Go out, close the door, and tell me what time it is,' I will ask you and say, 'No no no, say it again, there is some noise'"—she made a funny buzzing sound with her lips, *Brrrrrrrrip!*—"'between us.'" A great deal of care and effort goes into correcting for this noise, to ensure that the message heard by the B.I.P.M. accurately reflects the relative behavior of the world's clocks.

"We have eighty laboratories around the world," Arias said; some nations have more than one. "We need to organize all those times." She sounded gentle and encouraging, like Julia Child describing the essence of a good vichyssoise. First, Arias's team in Paris gathers all the necessary ingredients: the nanosecond-scale differences between each member clock and every other one, plus a strong dash of local data about the historical behavior of each clock. The information is then run through what Arias called "the algorithm," which takes into account the number of clocks in service (on any given day some clocks may be down for repair or recalibration), gives slight statistical favor to the more accurate of these clocks, and whisks the whole to a uniform texture.

The process is not purely computational. A human is needed to consider small yet critical factors: that not all labs calculate their clock data exactly the same way; that a particular clock has been behaving oddly of late and its contribution needs to be reweighted; that, owing to software errors, some of the minus signs in the spreadsheet were accidentally changed to plus signs and need to be changed back. Wielding the algorithm also involves a certain amount of individual, mathematical artistry. "There is some personal flair involved," Arias said.

The final result is what Arias calls "an average clock," in the best sense: its time is more robust than any single clock or national ensemble could hope to provide. By definition and by universal agreement, or at least by agreement of the fifty-eight signatory countries, its time is perfect.

It takes time to make Coordinated Universal Time. Simply ironing out the uncertainty and noise from all the G.P.S. receivers takes two or three days. The task of calculating U.T.C. would be logistically overwhelming if it were done continuously, so each member clock takes a reading of local time every five days at exactly zero hour U.T.C. On the fourth or fifth day of the following month, each lab sends its accumulated data to the B.I.P.M. for Arias and her team to analyze, average, check, and publish.

"We try to do it as soon as possible, without neglecting any checking," she said. "That process takes more or less five days. We receive on the fourth or fifth of the month, start calculating on the seventh, publish on the eighth or ninth or tenth." Technically, what is being assembled is International Atomic Time; creating U.T.C. is a simple matter of adding on the correct number of leap seconds. "Of course there is no clock providing U.T.C. exactly," Arias said. "You only have local realizations of U.T.C."

I suddenly understood: the world clock exists only on paper and only in retrospect. Arias smiled. "When people say, 'Can I see the best clock in the world?' I say, 'Okay, here you are, this is the best clock in the world.'" She handed me a sheaf of papers stapled in one corner. It was a monthly report, or circular, that is distributed to all the member time laboratories. The report, called *Circular T*, is the main purpose and product of the B.I.P.M. Time Department. "It is published once every month, and it is giving information on time in the past, which is the month before."

The world's best clock is a newsletter. I flipped through its pages and saw column after column of numbers. Listed down the left were the names of the member clocks: IGMA (Buenos Aires), INPL (Jerusalem), IT (Torino), and the rest. The columns across the top were dated every five days through the previous month—Nov. 30, Dec. 5, Dec. 10, and so on. The number in each cell represented the difference between Coordinated Universal Time and the local realization of U.T.C. as measured by a particular laboratory on a particular day. On December 20th, for instance, the figure for the national clock of Hong Kong was 98.4, indicating that, as of that moment of measurement, the national clock of Hong Kong was 98.4 nanoseconds behind Coordinated Universal Time. In contrast, the figure for Bucharest's clock that day was minus 1118.5, indicating that it was 1118.5 nanoseconds—a sizable step— ahead of the universal average.

The purpose of *Circular T*, Arias said, is to help member laboratories monitor and refine their accuracy relative to U.T.C., a procedure known as "steering." By learning how far their clocks deviated from the U.T.C. average during the previous month, member labs can tweak and correct their equipment to perhaps aim a little closer next month. No clock ever achieves perfect accuracy; consistency is sufficient. "It is useful because laboratories pilot their U.T.C.s," Arias said; she made time sound like a ship in a channel. "They need to know how the U.T.C. locally behaves. So they check if they have correctly steered to *Circular T*. That's why they're all checking their email and the Internet, to know where they were last month with respect to U.T.C."

For the most accurate clocks, steering is essential. "Sometimes you have a very good clock, and then it takes a time step—a jump in time," Arias said. On her copy of the latest *Circular T*, she pointed to the row of numbers representing the U.S. Naval Observatory. Its figures were all admirably small, in the range of double-digit nanoseconds. "This is an excellent realization of U.T.C.," Arias said. That's no surprise, she added, since the U.S. Naval Observatory, which has the largest number of clocks in the international pool, represents roughly twenty-five per-cent of the total weight of U.T.C. The U.S. Naval Observatory is respon-sible for steering the time utilized by the G.P.S. satellite system, so it has a global responsibility to follow U.T.C. very strictly.

But steering isn't for everyone. Piloting one's clock requires expen-sive equipment, and not all laboratories can afford to bother. "They let their clocks live their life," Arias said. She noted a row of numbers from a laboratory in Belarus, which seemed to be living a life of leisure, well off the standard. I asked whether the B.I.P.M. ever rejects a laboratory's contribution as too inaccurate. "Never," Arias replied. "We always want their time." As long as a national time lab is equipped with a decent clock and receiver, its contributions are averaged in to U.T.C. "When you build time," she said, one of the goals is "the broad dissemination of time"—U.T.C. can't be considered universal unless it includes everyone, no matter how out of step they might be.

I was still wrapping my head around what, and when, Coordinated Universal Time is. ("It took me a couple of years," Tom Parker later told me.) To the extent that a paper clock can be said to exist, it does so only in the past tense, derived as it is from data gathered the previous month; Arias calls U.T.C. "a post-real time process," a dynamic preterit. Then again, the numbers in the columns of her paper clock serve much like course corrections or channel markers for the real clocks out there, to help them steer in the right direction—as if U.T.C. were a future noun, like a harbor just over the horizon. When you look to your watch, clock, or cell phone for a reading of the official time, as derived from Boulder or Tokyo or Berlin, what you receive is only a very near estimate of the correct time, which won't be known for another month or so. Perfectly synchronized time evidently does exist—just not anymore and not quite yet; it is in a perpetual state of becoming.

I had come to Paris under the assumption that the world's most exact time emanates from some tangible, ultrasophisticated device: a fancy clock with a face and hands, a bank of computers, a tiny, shimmering rubidium fountain. The reality was far more human: the world's best time—Coordinated Universal Time—is produced by a committee. The committee relies on advanced computers and algorithms and the input of atomic clocks, but the metacalculations, the slight favoring of one clock's input over another's, is ultimately filtered through the conversation of thoughtful scientists. Time is a group of people talking.

Arias noted that her Time Department operates within a still-larger ensemble of consultative committees, advisory teams, ad hoc study groups, and monitoring panels. It hosts regular visits from international experts, holds occasional meetings, issues reports, and analyzes the feedback. It is checked, supervised, calibrated. Occasionally the over-arching Consultative Committee for Time and Frequency, or C.C.T.F., weighs in. "We don't operate alone in the world," she said. "For minor things we can make decisions ourselves. For major things we have to submit proposals to the C.C.T.F., and the experts from the best laboratories will say, 'We agree' or 'We don't agree.'"

All this redundancy is designed to counterbalance one ineluctable fact: no single clock, no single committee, no individual alone keeps perfect time. That's the nature of time everywhere, it turns out. As I began talking with scientists who study how time works in the body and mind, they all described its operation as some version of a congress. Clocks are distributed throughout our organs and cells, working to communicate and keep in step with one another. Our sense of time's passage is rooted not in one region of the brain but results from the combined working of memory, attention, emotion, and other cerebral activities that can't be singularly localized. Time in the brain, like time outside it, is a collective activity. Still, we're accustomed to imagining an ultimate collective somewhere in there—a core group of sifters and sorters, like an internal Bureau International des Poids et Mesures, perhaps run by a brown-haired Argentine astronomer. Where is the Dr. Arias in us?

At one point I asked Arias to describe her personal relationship to time.

"Very bad," she replied. There was a small digital clock on her desk; she picked it up and aimed its readout at me. "What time is it?"

I read the numbers. "One-fifteen," I said.

She motioned for me to look at my wristwatch: "What time is it?"

The hands read 12:55 p.m. Arias's clock was twenty minutes fast.

"At home, I don't have two clocks giving the same time," she said. "I am very often late for appointments. My alarm clock is fifteen minutes in advance."

I was relieved to hear this but I was troubled on behalf of the world. "Maybe that's what happens when you think about time all the time," I offered. If it's your job to coordinate the world's clocks, to create from Earth's gradients of light and dark a uniform and unified time, maybe you look to home as your refuge, the one place where you can ignore your watch, kick off your shoes, and enjoy some truly private time.

"I don't know," Arias said, with a Parisian shrug. "I have never missed a flight or missed a train. But when I know I can take this little degree of freedom, I do."

We commonly talk about time as an opponent: thief, oppressor, master. In a 1987 book called *Time Wars*, written at the start of the digital age, the social activist Jeremy Rifkin lamented that humanity had

embraced "an artificial time environment" ruled by "mechanical con-
trivances and electronic impulses: a time plane that is quantitative,
fast-paced, efficient, and predictable." Rifkin was particularly troubled
by computers because they traffic in nanoseconds, "a speed beyond the
realm of consciousness." This new "computime," as he called it, "represents the final abstraction of time and its complete separation from
human experience and the rhythms of nature." In contrast he praised
the efforts of "time rebels"—a broad category that included advocates
of alternative education, sustainable agriculture, animal rights, women's
rights, and disarmament—who "argue that the artificial time worlds we
have created only increase our separation from the rhythms of nature."
Time, in this telling, is a tool of the establishment and an enemy of both
nature and self.

The rhetoric is excessive but thirty years later Rifkin's complaint
does strike a common chord. Why else are we obsessed with productivity and time management if not to discover some saner way of navigating our lives? It's not "computime" that haunts us as much as our
slavish attachment to handheld computers and corporate-branded
smartphones, which allow the workday and workweek to never end.
Not wearing a watch was my way of shrugging off The Man, even if I'd
never laid eyes on him.

Still, to cast blame on "artificial" time is to give nature too much
credit. Maybe there was a time when time was a strictly personal affair,
but it's hard to imagine how long ago that would have been. Medieval
serfs toiled to the distant sound of village bells; centuries earlier, monks
rose, chanted, and prostrated themselves to the rhythm of chimes. In
the second century B.C.E., the Roman playwright Plautus rued the
popularity of sundials, which "cut and hack my days so wretchedly into
small pieces." The ancient Incas used a complex calendar to calculate
when to sow and harvest and to identify the most auspicious times for a
human sacrifice. (The calendar included a recurring "Vague Year" with
eighteen months of twenty days each plus, at the end, five "nameless
days" of ill omen.) Even early humans must have taken note of the daylight on the cave wall, in order to hunt effectively and return safely before dark. Even if any one of these customs were closer than today's to

"the rhythms of nature," it would be hard to embrace as a model that Earth's several billion residents should follow.

I looked again at the sheaf of papers Arias had handed me, then at her clock, then at my watch: it was time to go. For months I'd been reading the works of sociologists and anthropologists arguing that time is a "social construct." I'd interpreted the phrase to mean something like "artificially flavored," but now I understood: time is a social phenomenon. This property is not incidental to time; it is its essence. Time, equally in single cells as in their human conglomerates, is the engine of interaction. A single clock works only as long as it refers, sooner or later, obviously or not, to the other clocks around it. One can rage about it, and we do. But without a clock and the dais of time, we each rage in silence, alone.

THE DAYS

Thus this never-ending day began. To describe it all would be tedious. Nothing really happened; and yet, no day in my life was more momentous. I lived a thousand years, and all of them were agonizing. I won a little and lost a lot. At the day's end—if it can be said to have had an end—all that I could say was that I was still alive. Granting the conditions, I had no right to expect more.

—Admiral Richard Byrd, *Alone*

When I wake at night I'm tempted to look at the clock, but I already know what time it is. It's the same time it always is when I wake at this hour: 4 a.m., or 4:10 a.m., or once, for a disconcerting stretch of nights, exactly 4:27 a.m. Even without looking I could deduce the time, from the ping of the bedroom radiator gathering steam in winter or the infrequency of the cars passing by on the street outside. "When a man is asleep, he has in a circle round him the chain of the hours, the sequence of the years, the order of the heavenly bodies," Proust wrote. "Instinctively he consults them when he awakes, and in an instant reads off his own position on the earth's surface and the time that has elapsed during his slumbers."

We do this at all hours, knowingly or not. Psychologists call it temporal orientation and it's the hallmark of what one might call an adult sense of time: the ability to know the time, day, or year without turning to a clock or calendar. Numerous studies have attempted to understand how we achieve this orientation. In one experiment, researchers stood on the street and asked passersby a simple question—"What day is today?"—or presented a true-or-false statement ("Today is Tuesday") and noted the response. They found that people correctly name the current day more quickly if they are asked on or near the weekend. Some work out the answer by thinking retrospectively—"Yesterday was X, so today must be Y"—while others work backward from tomorrow. Which direction they orient toward depends on which weekend is nearest, the one past or the one approaching. You are more likely to calculate "today" based on yesterday if today is Monday or Tuesday; closer to Friday, your reference point shifts toward tomorrow.

Perhaps we place ourselves by means of temporal landmarks: we orient to the weekend, as if to an island on the horizon ahead or behind, and approximate our location on the sea of days. (For that matter, it's notable how often we talk about time in terms of space: next year is still "far away," the nineteenth century is the "distant" past, my birthday is "coming right up," like an approaching station.) Or perhaps we internally compile a list of the days that today might be and cross off the ineligible candidates until only one remains. ("It might be Thursday, but it's definitely not Wednesday, because I always go the gym on Wednesday mornings and I'm not carrying my gym bag.") Neither model quite explains why our temporal reference point shifts midweek—why our backward-looking thoughts decline as the week advances. By whatever method, we engage in such orienteering virtually nonstop, across seconds, minutes, days, and years. We wake from a dream, exit a movie, look up from an absorbing book and think, Where am I? When am I? We lose track of time and need a moment to right ourselves again.

That I can know, without looking, what time it is when I wake in the middle of the night may also be a simple matter of induction: it was 4:27 a.m. the last time I woke in the night as well as the time before that, so it's probably 4:27 a.m. now. The question is why, or how, I can be so consistent. William James wrote, "All my life I have been struck by the accuracy with which I will wake at the same *exact minute* night after night and morning after morning, if only the habit fortuitously begins." In that moment, of all my waking moments, I am very aware of being at the service of something; there is a machine in me, or I am a ghost in it.

In either case, once the ghost gets thinking, there is much to think about—most prominently, how little time I have in which to do all the things I'm thinking about and how behind I am already. "I see your book scheduled on my calendar," my editor writes. "I need to know where things stand." I began this project a few weeks before Susan was due to deliver twins, our first and only children. In retrospect the choice of timing was not ideal. Friends and family joked all too eagerly that if I was struggling to manage my time, not to worry, my kids would soon manage it for me.

Yet as fraught as these waking moments are, they are also calming— even expansive—and it feels to me that occupying them is like being inside an egg. That idea comes to me one night right before I go to sleep; I make a note of it in my bedside notebook and am surprised and delighted later, at (I assume) 4:27 a.m., to find myself inhabiting the very notion. It's as if, in falling asleep, I've fallen into that same egg and woken as pure yolk, cushioned and aloft on an extended present. It won't last, I know. In the morning, the hours and minutes will reassert themselves and this seemingly limitless expanse of time will have evaporated or been locked away beyond reach; I'll be outside the shell trying to imagine my way back in. That's the fundamental tension of modern life: the dream of boundless time, dreamt from the confines of an egg carton. But that's a thought for tomorrow. Now, my bedside clock ticks, like the distant clicking of a kitchen egg timer or the muffled beat of a heart.

Once upon a time, a man entered a cave and stayed there alone for many days and nights. He saw no natural light; no sunrise or sunset arrived to announce when a day officially began or ended; he had no clock or watch to mark the passage of his moments and hours. He wrote; he read Plato; he thought a lot about his future. He was alone with time for a very long time, although not quite the length of time he expected.

That was Michel Siffre's first temporal experiment, in 1962. Siffre, a twenty-three-year-old French geologist, had recently discovered an underground glacier, the Scarasson, in a cavern in southern France. The Cold War and Space Race were on; fallout shelters and space capsules were much in discussion. Like many scientists, Siffre wondered how a human would manage in such places, isolated from other people and from the sun. His initial idea was to spend two weeks studying the cavern. But he soon determined to stay longer, for two months, to explore what he later called "the idea of my life." He would live "like an animal," he told *Cabinet* magazine in 2008, "in the dark, without knowing the time."

He pitched a tent, with a sleeping bag on a cot. He slept, rose, and ate as he wished and kept a written record of his activities; a small generator powered a lamp by which he read, studied the glacier, and moved about. He was cold and his feet were perpetually wet. His only contact with the surface was by telephone, and he regularly called his colleagues above—who were under strict instruction not to betray any information about the day or time—to report his pulse rate and his proceedings.

Siffre entered the cave on July 16th and planned to leave on September 14th. But on August 20th, by his calendar, his colleagues called to say that his stay had ended; his time was up. By his reckoning only thirty-five days had passed—thirty-five days of waking, sleeping, and puttering—yet by the outside clock, sixty days had elapsed. Time had flown.

By accident, Siffre had discovered something important about human biology. Scientists were already aware that plants and animals have an innate ability to track a roughly twenty-four-hour period—a circadian cycle. (The word comes from the Latin phrase *circa diem*, "about a day.") In 1729 the French astronomer Jean-Jacques d'Ortous de Mairan noted that a heliotrope plant, which opens its leaves at dawn and closes them at dusk, continued the behavior even when kept in a dark closet; it seemed to innately grasp when day and night had occurred. For camouflage, fiddler crabs change their color on a fixed schedule over the course of the day, from gray to black and back again, even in the absence of daylight. Light-deprived fruit flies emerge from their pupal cases religiously at dawn, a time when the air is at its most humid, an adaptation that prevents novice wings from drying out. This internal, circadian rhythm doesn't precisely match the external rhythm of daylight and darkness; the circadian clock runs a little longer than twenty-four hours in some organisms, a little shorter in others. A heliotrope kept in the dark for too long will eventually fall out of step with the natural cycle of the day; it's not very different from my wristwatch, which, unconnected to the radio and satellite signals that disseminate perfect world time, requires that I reset it daily.

By the nineteen-fifties it was clear that humans also have an endogenous circadian clock. In 1963 Jürgen Aschoff, the head of the Biological Rhythms and Behavior department at what was then the Max Planck Institute for Behavioral Physiology, in West Germany, converted a soundproof bunker into an experimental station where subjects would stay for weeks, without mechanical clocks, as he monitored their physiology. Siffre's experiment in the Scarasson was among the first to show that our circadian cycle is not precisely twenty-four hours long. The period that Siffre was awake each day varied greatly in length, from as little as six hours to as many as forty, but on average he settled into a sleep-wake cycle that was twenty-four hours and thirty minutes long. This soon put him out of sync with the surface day, and the experience—that of an animal trapped alone with the idea of his life—unsettled him. He had descended with the aim of studying the effect of extreme isolation on the human psyche; he emerged as an unwitting pioneer of human chronobiology and, he later recalled, as "a half-crazed, disjointed marionette."

The most commonly used noun in American English is *time*. But if you ask a scientist who studies time to explain what time is, he or she invariably will turn the question on you: "What do you mean by time?"

And already you've learned something. You might begin, as I did, by qualifying your statement to mean "time perception," to distinguish between external time and your internal grasp of it. This dichotomy suggests a hierarchy of truth. Foremost is time as told by one's wristwatch or the clock on the wall, which we typically think of as "true time" or "the actual time." Then follows our perception of this time, which is accurate or not depending on how closely it matches the mechanical clock. I've come to think that this dichotomy is, if not meaningless, certainly of little help in trying to understand on a human scale where time comes from and where it goes.

But I'm jumping ahead. One of the oldest debates in the scientific literature is whether "time" is something that can be "perceived" at all. Most psychologists and neuroscientists have come around to thinking that it isn't. Our five senses—taste, touch, smell, sight, and hearing—all involve discrete organs that detect discrete phenomena: sound is what we call it when vibrating air molecules trigger movements of the tympanum in the inner ear; sight is what results when photons of light strike specialized nerve cells at the back of the eye. In contrast, the human body has no single organ devoted to sensing time. The average person can sense the difference between a sound lasting three seconds and

another sound lasting five seconds, as can dogs, rats, and most laboratory animals. Yet scientists still struggle to explain how the animal brain
tracks and measures time on so fine a scale.

One key to understanding what time is, physiologically, is to know
that when we talk about time, we may be referring to any of a number of
distinct experiences, including

> *Duration*—the ability to determine how much time has elapsed
> between two specific events or to accurately estimate when
> the next event will occur.
> *Temporal order*—the ability to discern the sequence in which
> events occurred.
> *Tense*—the ability to discriminate between the past, present, and
> future, and the understanding that tomorrow lies in a different
> temporal direction than yesterday.
> *The "feeling of nowness"*—the subjective sense of time passing
> through us "right now," whatever that is.

Suffice to say, discussions of time often get confusing because we're
using just one word to describe a multilayered experience; to the scientific connoisseur, *time* is as generic a noun as *wine*. Many of these temporal experiences—duration, tense, simultaneity—feel so basic and
innate that they hardly seem to merit distinction. But that's only true
from an adult perspective. The view in developmental psychology is
that time is something that humans come to know only gradually. One
fundamental insight comes to us in the first few months of life, when
we learn to distinguish "now" from "not now"—although the seeds of
this awareness probably reach us even sooner, while we're still in the
womb. Not until age four or so can children accurately distinguish "before" from "after." And as we age, we become ever more keenly aware of
the "arrow of time" and its unidirectional flight path. Our knowledge of
time is hardly as a priori as Kant proposed. Not only is time something
that gets in us, it takes years to fully do so.

We think about time constantly: we estimate its duration, consider yesterday and tomorrow, distinguish before from after. We dwell in time and on it, anticipating, remembering, remarking on its passage. By and large these are conscious experiences and, so far as we can tell, unique to our species. But underneath, requiring no thought, infusing all life going back nearly four billion years, is the circadian cycle, the time of days. For a biological phenomenon it is remarkably mechanical in its reliability, and in the past two decades scientists have made great strides in delineating its genetic and biochemical underpinnings. Of all the clocks in us, the circadian clock is by far the best understood. If the scientific exploration of human time were to be mapped as a physical journey, it would begin on solid ground and in daylight, with our knowledge of circadian rhythms, and descend into a marshy dusk.

Circadian rhythms are commonly associated with one's sleep-wake cycle. But this is a misleading indicator: although your sleep patterns are influenced by your circadian clock, they are also subject to conscious control. You can elect to be early to bed and early to rise; to live like an owl, sleeping all day and awake all night; or even to forgo sleep for days on end. The circadian clock isn't overridden so easily; if it were, it wouldn't be worth counting on.

A more accurate way to track circadian rhythms, at least in humans, is through body temperature. Although it's often said that the average human body temperature is 98.6 degrees (actually it's 98.4), that's only an average. Over the course of a day, your temperature varies by as much as two degrees; it oscillates, peaking in mid- to late afternoon and dropping to its low point in the predawn hours before you wake. We differ individually in the exact amplitude of this peak temperature and in its timing; activity or sickness can warm the body too. But we all express a clockwork rise and fall in body temperature through each day, day after day.

Other bodily functions obey a strict circadian cycle too. Your resting heartbeat can vary by two dozen beats per minute depending on the time of day. Blood pressure oscillates across twenty-four hours; it is lowest between two and four o'clock in the morning, rises through the day, and peaks at around noon. We pass less urine at night than during the day, not only because we drink less then but because the activity of hormones (the release of which also follows a circadian cycle) makes the kidneys retain more water. You could schedule your daily tasks around the circadian clock. Physical coordination and re-action times peak in midafternoon; the heart is most efficient and the muscles strongest at around five or six in the afternoon; one's threshold for pain is highest in early morning, making it the ideal time for dental surgery. Alcohol is metabolized most slowly between ten at night and eight in the morning; the same drink lingers longer in your system at night than during the day, so it makes you more drunk. Your skin cells divide most rapidly between midnight and four in the morning, whereas facial hair grows faster during the day than overnight. A man who shaves in the evening does not wake up with five-o'clock-in-the-morning shadow.

These rhythms strongly influence one's health. Strokes and heart attacks are most common in late morning, when blood pressure rises most rapidly. Because hormone levels oscillate naturally over twenty-four hours, the efficacy of drugs varies greatly too, depending on what time of day they are administered, a fact to which doctors and hospitals are increasingly attuned. The same holds true for animals of all kinds. In one unfortunate lab study, a potentially fatal dose of adrenaline killed as few as six percent, and as many as seventy-eight percent, of the rats that received it, depending on when it was given. Certain insecticides kill more of their target insect in the afternoon. Circadian rhythms also affect one's mood and mental agility. One study asked subjects to cross out as many instances of the letter *e* in a magazine as they could manage in thirty minutes; they did worst at eight in the morning and best at eight-thirty at night.

Alertness is strongly circadian too; it peaks when body temperature peaks and it bottoms out when body temperature is lowest. The latter

28

period occurs, for most people, in the hours before dawn. One result is that night-shift workers aren't nearly as productive as they may think they are. Between three and five in the morning, workers are slowest to respond to a warning signal and most likely to misread a meter. The mathematician Steven Strogatz has noted that the accidents at Chernobyl, Bhopal, Three Mile Island, and aboard the *Exxon Valdez*, which have all been attributed to human error, occurred in these hours, which shift workers call the zombie zone. As a species, we humans are curiously willing to entertain this zombie, and increasingly our science is revealing the ill effects of this flirtation.

A clock is a thing that ticks. The tick can be almost anything as long as it's persistent and steady: the vibrations of atoms, a swinging weight, a planet turning on its axis or orbiting the sun. A mere lump of coal ticks. Coal is made of carbon atoms, which ordinarily have six protons and six neutrons (in carbon-12), although one in a trillion or so contains six protons and eight neutrons (carbon-14). The ratio of carbon-14 to carbon-12 is fairly constant in living things but it declines the longer something has been dead, because the carbon-14 atoms gradually decay into nitrogen-14. This happens, on average, every fifty-seven hundred years. Equipped with a knowledge of this decay rate and the ratio of carbon-14 to carbon-12 in your lump of coal, you can calculate the coal's age, in tens of thousands of years. Coal, or any carbon fossil, is a clock of eons.

Whether a clock—a planet, a pendulum, an atom, a rock—also counts the ticks is a matter of long philosophical debate. A sundial tracks a moving shadow around its face; the hours are marked with printed numerals. Does the clock count the numbers or do you? Does time exist independently of the mind that counts it? "Whether, if soul did not exist, time would exist or not, is a question that may fairly be asked," Aristotle mused, "for if there cannot be someone to count, there cannot be anything that can be counted." It's like the koan about the falling tree in the forest: is coal a clock if there is no scientist to measure its C-14/C-12 ratio? Augustine was resolute: time resides in the measuring of it, which makes it solely a property of the human mind. One hears an echo of Augustine in the late physicist Richard Feynman, who pointed out that the dictionary definition of time is circular: time is a period, which is defined as a length of time. Feynman added, "What really matters anyway is not how we define time, but how we measure it."

In the circadian clock, what ticks are a cell's contents—genes, proteins—and the dialogue among them. Every living cell contains DNA, tightly coiled strands of genetic material. In eukaryotes, a broad group of organisms that includes all animals and plants, DNA is kept within the membrane of a cell nucleus. Each strand of DNA is actually two strands joined zipper-like down the middle, forming a double helix. The strands, in turn, are made up of nucleotides, which form genes of varying lengths. DNA is intensely dynamic. It regularly unzips to expose a gene (or a few), makes a working copy, and sends this out of the nucleus into the cell's cytoplasm, where different kinds of proteins are constructed based on the incoming template. Picture a busy architect on an island mailing a blueprint to a manufacturer on the mainland, who will use it to build various robots there.

Most genes code for proteins that perform activities elsewhere in the cell; these assemble into molecules, catalyze metabolic reactions, repair internal damage. But the circadian-clock genes—there are two main ones—are different. These code for a pair of proteins that accumulate in the cytoplasm and eventually seep back into the nucleus, where they attach to the activators of the original genes and shut them down. In short, the "clock" is little more than a pair of genes that, eventually and by means of various intermediaries, turn themselves off. Our architect isn't simply mailing out a blueprint; she is sending out bottled messages addressed to her future self. Eventually, when enough bottles accumulate in the sea, the message will reach her, and the message says, "Take a nap."

When the architect falls asleep and the clock genes are at rest, protein production ceases. The existing proteins degrade in the cytoplasm and stop pressing into the nucleus and shutting off the genes, freeing them to again issue orders. If that process sounds circular, that seems to be what natural selection was favoring. What's remarkable isn't what is produced (which is nothing physical, all told) but the period of production: the cycle from the moment the clock genes are first activated, then switched off, then switched on again takes, on average, twenty-four hours. Something is produced after all: not a molecule but an interval. At heart, the circadian clock is a conversation—between a cell's DNA and its protein builders—that takes about a day to unfold. This endogenous clock will tick through its cycle even if its bearer—a person, a mouse, a fruit fly, a flower—is kept in darkness for days on end. But because it isn't exactly as long as the daylight cycle, it will fall increasingly out of step with the solar day; regular exposure to sunlight resets the circadian clock and keeps it aligned. Sunlight is the conversation's moderator, weighing in daily to keep it on track but not intervening at every moment.

That the period of this clock is about twenty-four hours is all the more remarkable given that most biochemical reactions in a cell occur in just fractions of a second. In practice, the dialogue between the clock genes in the nucleus and the proteins in the cytoplasm is mediated by an array of additional molecules, which are encoded by genes of their own. It's less a conversation, perhaps, than a crazy game of telephone. Our architect sends a message to herself, but there are middlemen to contend with—contractors, delivery boys, doormen. Finally, her note has arrived: twenty-four hours!

Much of what scientists know about the circadian clock has been gleaned from animal studies. In the nineteen-sixties a classic series of experiments by Seymour Benzer and Ronald Konopka revealed that fruit flies become more and less active according to a regular twenty-four-hour cycle. Moreover, certain strains of fly exhibited rhythms that were slightly, and sometimes drastically, longer or shorter than twenty-four hours. By crossbreeding flies and tweaking the DNA, biologists identified the genes involved and revealed a basic model for how the clock works. A pair of genes, nicknamed *per* and *tim* (for "period" and "timeless"), code for the production of a pair of proteins called PER and TIM. The two proteins join to form a single molecule; when enough of this molecule has accumulated in the cytoplasm, it leaches back into the nucleus and works to turn off the genes *per* and *tim*.

Subsequent studies have revealed a very similar clock, involving very similar components, in mice, although the mouse clock has additional variations of the key genes and proteins. The same genetic components have been identified in human cells too. Indeed all animals, from ants and bees to reindeer and rhinoceroses, run on a circadian clock of similar construction. Plants have a circadian clock, which many species use to turn on their chemical defenses in anticipation of morning attacks from insects; the plants are more resistant to attack when their clocks are functioning normally. Janet Braam, a cell biologist at Rice University, and her colleagues found that the circadian clocks of cabbages, blueberries, and other fruits and vegetables continue to tick even after the plants have been harvested. But under the constant light of a grocery store—or the constant dark of a refrigerator—the circadian rhythms start to dissipate, as does the cyclical production of key compounds, making the plant more susceptible to bugs and perhaps diminishing its taste and even its nutritional value. We are turning our vegetables into . . . vegetables.

Even the lowly but well-studied *Neurospora*, a mold that grows on bread, runs on a circadian clock. The common aspects of the plant and animal clocks are sufficiently striking and deep-seated that some biologists suspect that we all have been running on some version of the same clock ever since the first multicellular organisms appeared on Earth seven hundred million years ago. I find this idea comforting at 4:27 a.m. as I consider my consciousness and mortality. I'm a member of perhaps the only species that anticipates an end. The grass prepares to receive sunlight, untroubled by the prospect of my mower. As I rise, so rise the bees, the blossom on the distant plant that will produce the coffee for my coffeemaker, the mold accumulating on the bread on my kitchen counter. The same inheritance ticks in us, telling us the time and leaving those who can to count it.

· · ·

We want to know what time it is. We ask the bedside clock; we ask the watch; we ask one another: *Would you tell me the time?*

All's well until we ask a second clock, which invariably disagrees with the first. Which one to believe? So we find another clock to do the mediating: the tower clock in the town square, the punch clock by the foreman's door, the clock on the wall of the principal's office that rings the bell when the school day ends. For any one of us to be on time, we must all agree on what that one time is so that we can all be on it together, at the same time. We must be synchronized. Life is one great adaptation to the when of others.

This is equally true for our cells. In the nineteen-seventies it became clear that the main circadian clock in mammals is a brain structure called the suprachiasmatic nucleus, a dual cluster of some twenty thousand specialized neurons in the hypothalamus, near the base of the brain, that fire in unison and with a circadian rhythm. It takes its name from the fact that it sits just above the optic chiasma, where the optic nerves from the right and left eyes cross (a handy location for receiving information about the outside world), and it regulates the daily rise and fall of the body's temperature, blood pressure, rate of cell division, and other vital activities. It is reset by daylight but also marches to its own beat; left alone in a dark cave or bathed in constant light, it would repeat its rhythm on average every 24.2 hours, almost but not quite in step with the twenty-four-hour rhythm of day and night. If the structure is removed from a lab rodent or squirrel monkey, the animal becomes asynchronous. Its body temperature, its release of hormones, its physical activity show no circadian pattern, and without a shared clock these processes no longer stay in step with one another. Hamsters in this state become diabetic and can't sleep; they are disoriented and their movements become disjointed. But when suprachiasmatic nucleus cells are transplanted back in, the animal regains its clock—albeit the clock of the donor.

But that cluster of cells isn't the only clock we contain; in the past decade it has become clear that virtually every cell in the human body contains a circadian clock of its own. Muscle cells, fat cells, cells of the

pancreas, liver, lungs, and heart, even whole organs keep their own circadian time. A study of twenty-five kidney transplant patients found
that, in seven of them, the new kidney ignored the new owner's circadian rhythm and instead stuck to the excretory rhythm it had expressed
in the donor. The other eighteen kidneys fell into sync with the new
owners' internal rhythm but in opposition: they were most active when
the existing kidney was least active and vice versa. Even genes, which
produce proteins, maintain our cells, manage our internal energy grid,
and ultimately help define who we are, function on a circadian schedule.
Until about a decade ago only a low percentage of mammalian genes
were thought to oscillate with circadian regularity, but that rhythm now
appears to be a basic property of all genes. We are filled with clocks,
trillions upon trillions of them.

Each of these clocks is potentially autonomous; it ticks on its own
and, if isolated from the rest, will free-run on a near-daily rhythm. Moreover, few of these clocks oscillate in precisely the same phase; a study
of more than a thousand genes in the heart and liver of mice found that
their activity varied on a circadian schedule but not all as one. Picture
an orchestra: the string section—violins, violas, cellos, and basses—
express a multilayered theme; the brass and woodwinds enter with a
counterpoint; the percussion rumbles in the back and stands out occasionally with a gong. But without a conductor the result would all be
noise. In humans, and in many vertebrates, the conductor is the suprachiasmatic nucleus. It keeps the primary beat and conveys it, through
hormones and neurochemicals, to the peripheral clocks, keeping them
in step with one another. To be of any service, a clock must tell its time
to the clocks around it or at the very least listen and absorb what the
others have to say. A clock is a concert, a group conversation, an inter-
active story. You don't simply carry many clocks within you; the sum of
you *is* a clock.

But the whole-body clock isn't a perfect clock either, at least not by
itself. To stay synchronized to the twenty-four-hour cycle of day and
night, it must be reset, ideally daily, with input from the outside world.
By far the strongest cue is daylight, and in humans, as in all mammals
and most animals, the gateway to light is the eye; if the suprachiasmatic

nucleus is the body's conductor, the eye is its metronome, translating physical time into something that physiology can understand. A discrete neural pathway, called the retinohypothalamic tract, runs from the back of the eye into the suprachiasmatic nucleus; when daylight registers within the eye, signals travel to the body's conductor and prompt it to start the symphony again from the top.

This process, called entrainment, is essential if the body is to keep its many clocks working as a unit. The conductor can't be reset at any moment, by any light; over the years, scientists have learned a great deal about which wavelengths of light are the most effective, the optimal length of exposure, and at which hours of the day. In sleep labs, with special lighting devices, people can be reprogrammed to function on days of different lengths—twenty-six-hours, twenty-eight hours—or to wake at midnight and go to sleep at noon. Left to our own devices, however, we entrain to the day and the steady tick of Earth's rotation. My cell phone synchronizes with the rest of the world by sending a signal to an orbiting satellite that contains a superaccurate clock of its own, then waits for the reply. To synchronize my brain with the rest of the world, I need only open my eyes and let in the day.

Once upon a time, a cell entered a cave and stayed there for many days and nights. It was me; it was you. It was, in the months before their birth, Leo and Joshua, our fraternal twin sons.

Are we born into time or is time born into us? The answer depends on what one means by time, of course, but also what is meant by "we" and when this we begins. Start with a single cell: a living, semisealed factory of biochemical reactions and interactions, energy cascades, ion exchanges, feedback loops, and the rhythmic expressions of genes. The sum total of this activity can be measured as a subtle rise and fall of the cell's electrical potential over time. One cell becomes two, then thousands, then an identifiable embryo. Sometime between forty and sixty days after conception, the cells appear that will become the suprachiasmatic nucleus. They arise in one part of the nascent brain, drift off, and by sixteen weeks, midway through pregnancy, have settled in the hypothalamus. In baboons, whose fetal development is similar to that of humans, the cells of the suprachiasmatic nucleus have begun to oscillate on their own by the end of gestation; the cells' metabolic activity shifts between high and low gear over a period roughly twenty-four hours long. In the absence of daylight, something close to the rhythm of the day has emerged. The cells have gone circadian.

A human fetus displays clear signs of organized circadian activity even earlier in gestation—at about twenty weeks, a month after the suprachaismatic nucleus has settled into place. The heart rate, respiration rate, and the production of certain neural steroids all vary regularly over a twenty-four-hour period. But the fetus isn't adrift on endogenous time like some free-running French speleologist. Its circadian activity occurs in sync with the natural light-dark cycle outside the womb, despite the fact that the fetus is in darkness and its retinohypothalamic tract, the pathway through which news of daylight reaches the master clock nucleus, has yet to form. How did the day get in there?

By way of its mother. Amid the nutrients and substances that flow in through the placenta are two neurochemicals—the neurotransmitter dopamine and the hormone melatonin—that play a critical role in syncing the fetus's master clock to the external time of day. Receptors for these neurochemicals appear on the suprachiasmatic nucleus early in the structure's formation in utero. Often when I lie awake in the dark at night I like to imagine that life in the womb must be like this only better—no clock ticks there, no thoughts of ticking even flicker through; the fetus floats in a space beyond time, unhurried and innocent. But clearly this is fiction; an embryo is continuously bathed and infused with the correct time of day. It lives and grows on borrowed time.

What does a fetus gain from a secondhand knowledge of the day? One possible advantage, scientists think, comes in the initial days outside the womb. Mammals that live in burrows—moles, mice, ground squirrels—often aren't exposed to direct daylight for the first few days or weeks after birth. If newborn pups, when they finally emerge above ground, had to spend several additional days acclimating to the rhythms of daylight, they would be especially vulnerable to predators. Perhaps for them, and for humans too, the circadian experience in the womb provides a kind of jump start, a preparatory course for bright reality.

But a circadian clock is also essential to organizing the internal environment. An animal, even an embryonic one, is an assemblage of miniature circadian clocks—billions upon billions of them, in cells, genes, and developing organs, working roughly twenty-four hours a day at their designated tasks. Without a central clock—provided in utero by the mother and eventually by the individual's suprachiasmatic nucleus—these various systems would neither develop properly nor work in coordination with one other. If the stomach decides to eat at one o'clock but the gastric enzymes show up an hour late, digestion will be inefficient. The maternal clock provides the fetus with essential organization—"a state of internal temporal order," as one journal article puts it—until the individual's clock is up to the task. It also integrates the embryo's physi-ology with the mother's, enabling the two to eat, digest, and metabolize on the same schedule. After all, up until the moment of birth, the fetus is literally part of the mother, one more peripheral clock to be governed and steered.

The mother's circadian rhythm may also serve as an alarm clock for the fetus. Researchers have found that, for many mammals, the onset of labor has a circadian component. For instance, rats typically give birth during daylight hours—the equivalent of their nighttime—and in the laboratory the onset of labor can be shifted by shortening or length-ening the mother's exposure to light. Among women in the United States, most at-home births occur at night, between one o'clock and five o'clock in the morning. (In hospitals, however, babies are most often born on weekdays between eight and nine in the morning, presumably due to the increase in induced births and C-sections, which are typi-cally scheduled to optimize care from the staff.) Several animal stud-ies suggest that the fetus also plays an active role in the scheduling of labor. On the last day of gestation, the master clock in the fetal brain, which is already synchronized to the solar day, triggers the cascade of neurochemical signals that culminate in birth. The young clock, once dark and peripheral, announces its independence and prompts its own liberation into the world.

Leo and Joshua were born six-and-a-half weeks early and four minutes apart in the early hours of the Fourth of July. Newborns are strange creatures—shocked and squalling and coated white with vernix. Looking back, I can frankly say that, when our boys first emerged in the delivery room, what I saw were two half-crazed, disjointed mar-ionettes. Small wonder. For several months up until that moment, they had known time intimately; it was a neurochemical bath piped in through the placenta. Now here were two fresh humans searching desperately for the bedside clock—*What time is it?*—with no quick hope of finding it.

Of course their new clock, the universal clock, was glaring at them in the form of light. (Granted, this was hospital light at two o'clock in the morning, but in a few hours they would be exposed to the real thing.) When Michel Siffre first emerged from endogenous cave-time into daylight, he benefited from having a mature circadian system. Within just a few days of his return to civilization his sleep-wake cycle had returned to something close to normal and he was again in sync with his friends, family, and the wider world. The newborn, in contrast, emerges with a circadian clock that isn't yet fully operational. He is born in sync with his mother and then, for some weeks and in broad daylight, descends into temporal chaos, dragging his new family down with him.

That would explain much of what happened in those first few weeks, as best as I can remember them. We all slept so little and so irregularly that my working memory dissolved. I can recall watching *The French Connection* several times after midnight while bottle-feeding two infants but even now I couldn't tell you the plot; there was a man with a beard, a subway chase, Gene Hackman in a porkpie hat. Much like Siffre, I could hardly remember what I'd done the previous day, or how long ago the previous day had occurred, or whether the previous day had even ended yet. That entire period blurred into one long stretch of wakefulness and sleeplessness. When after many months Susan and I at last regained the ability to reflect, we found ourselves saying both "Time stood still" and "Time flew," and each statement felt equally true.

For the first three months or so of life, an infant sleeps sixteen to seventeen hours a day but not in a consolidated manner. Its periods of rest are distributed fairly evenly over a twenty-four-hour period: more during the day than at night, at first, then by week twelve more at night than during the day. The scattershot pattern is the result of poor internal communication. Although a baby is born with a working circadian clock in the hypothalamus, the neural and biochemical pathways that convey the rhythm throughout the brain and body aren't yet all connected. "The clock is ticking," Scott Rivkees, the chair of pediatrics at the University of Florida College of Medicine, told me. "But there can be a mismatch between what happens in the clock and the rest of the organism." It's as if the U.S. Naval Observatory were unable to send its time signals to the network of G.P.S. satellites, or if N.I.S.T. neglected to switch on its time-only radio channel: a baby's brain realizes the correct time of day but fails to properly disseminate it.

Not long ago, this mismatch was the subject of great clinical interest. In the late nineteen-nineties, Rivkees helped to identify the retinohypo-thalamic tract, the neural pathway that connects the eyes to the supra-chiasmatic nucleus, in premature infants and newborns. He also found that the channel is functional by late in gestation: it responds to light even in infants that are born several weeks too early. The discovery, and its implications, took Rivkees by surprise, he told me. Infants born prematurely are kept in a neonatal intensive-care unit until they are strong enough to go home. Well into the nineteen-nineties, the common practice in these units was to always keep the lights off; the womb is dark, so a preemie's hospital environment should be too, the reasoning went. Rivkees wondered if that reasoning was sound. An infant born prematurely immediately loses circadian input from its mother—information that is vital if nascent organs and physiological systems are to develop in sync with one another. But the preemie does have a functional retino-hypothalamic tract, so potentially it could absorb circadian information on its own. Rivkees suspected that hospitals, in trying to do the right thing, were depriving infants of essential temporal data.

He and his colleagues ran an experiment. A control group of neonates was kept in the typical N.I.C.U. environment, with constant dim lighting, for two weeks before their release from the hospital. A second group was exposed to a cycled regime: the lights were on from seven in the morning to seven at night and off the rest of the time. The infants from both groups went home with activity monitors on their ankles that continuously recorded slight changes in heart rate and breathing. The data revealed that, after the first week at home, babies from both groups had essentially the same sleep patterns. But the infants that had been exposed to cycled lighting in the hospital were twenty to thirty percent more active in the daytime than they were at night and their mothers were more engaged with them; the control group didn't show comparable patterns for another six to eight weeks. Early light, and an early sense of time, provided more than a health boost; it was essential to the chemistry that helps a new family bond.

Thanks in part to that research, neonatal units now typically use cycled lighting. And pediatricians typically recommend the use of blackout shades at home only from dusk to dawn, not during an infant's afternoon nap. But the myth of the timeless womb persists among parents, Rivkees said. When pediatric nurses make house visits, they commonly find newborns sleeping in rooms that are kept continuously dark or dim. "You'd think that kids are going home to bright airy rooms, but that's often not the case," he said. Even after birth a mother continues to imprint her circadian rhythm on her infant. Breast milk contains tryptophan, a molecule that when ingested is synthesized into melatonin, a neurochemical that induces sleep. Naturally, tryptophan is produced on the schedule of the mother's circadian clock; more is available in the breast at certain times of day than at others. Regularly scheduled feedings help to cement an infant's sleep cycle to the mother's as well as to the natural day, and several recent studies suggest that babies that are breast-fed adopt a sane sleep schedule sooner than formula-fed babies do. To the newborn, the day is something to consume as well as to absorb.

I wake in the dark to a cry. It's Leo, hungry. What time is it? I fumble for the clock and bring it close to my eyes: 4:20 a.m. Today is June 21st, the first day of summer, the day with the most minutes of daylight. Evidently I will be awake for all of them.

With the aid of some twenty thousand clock cells and some specialized neurons in their retinas, Leo and Joshua have metabolized the daylight of almost their first three hundred and sixty-five days. For some weeks now they've slept through the night, but they rise painfully early, at the first whisper of dawn, even before the birds. Friends contend that if we put them to bed a little later than we normally do they'll wake a little later in the morning. But we've been reading up on circadian entrainment and are entrusting our sanity to science.

Light will reset the circadian clock, but not just any light will do;

otherwise the circadian clock would be reset with every passing moment of daylight. In practice, organisms are most sensitive to light—more precisely, to changes in the intensity of light—at the start of their day. The circadian clocks of nocturnal animals such as bats are more attuned to changes in the intensity of daylight in the evening than in the morning, while diurnal animals (including children, when they have achieved something like a diurnal stage) are more sensitive to the light at dawn than at dusk. So we could expect our kids to wake at the same early hour whether they went to bed at six o'clock or eight o'clock the night before.

No sooner have I discussed all this again in my head with Susan (she's awake now too) than the birds outside have burst into song: first a single, warbling robin, then a chorus. The time is 4:23 a.m. Susan shuffles off to feed Leo; in twenty minutes he's asleep again and she returns to bed. Less than a minute later, Joshua wakes with a squawk. A pale light seeps through the window blinds. The birdsong has become cacophonous and we think it's keeping Joshua awake. Scientists who study circadian rhythms use the term *zeitgeber*—from the German *Zeit* (time) and *Geber* (giver)—to characterize an event that resets the biological clock. The strongest and most common *zeitgeber* is daylight. Deprived of that input for long enough, humans will find other cues to unconsciously set their circadian rhythms to: an alarm clock, the ringing of a bell, even simple but regular social contact. Daylight is a *zeitgeber* to the robin, the robin is a *zeitgeber* to the child, the child is a *zeitgeber* to the man.

"Shut up, bird," Susan whispers.

It's becoming clear to the two of us that parenthood will be a gradual but relentless series of concessions. At first we told ourselves that we weren't actually new parents but managers of a startup company. According to this narrative, our lives were exactly the same as before except for the addition of two charming if underperforming employees. Our job was to impose on them a schedule—to eat at such-and-such times, to sleep from X o'clock to Y o'clock—that perfectly suited the schedule of our adult, childless, former selves. But increasingly our company appeared to be owned and operated by its alleged workers.

I became militantly attached to the boys' midday nap, a period of two or three hours when my former self could reassert itself and do former-self stuff, such as write or sleep, as if it were still my present self. But this was a fiction too. I'd ease both boys into their cribs and sneak away; they'd quiet down, but soon one would begin to chatter and call out for me. When I refused to visit, he would hoot and then start jumping up and down even as his brother slept soundly just a few feet away. This agitated me beyond all reason. It was an affront to my new, parental dictatorship and further eroded my sense of independence. *This is my time*, I tried to tell him.

I sweet-talked, cajoled, scolded. This only excited him, which deepened my resentment. When he wasn't frightened by my scowl, he seemed to be taking pleasure in needling me with his antics; with a start I realized that I had become The Man and now the boy was sticking it to me. Eventually I understood: he didn't really want to stick it to The Man, he just wanted The Man to play with him. I surrendered. I gave up the delusion of a workday, and the two of us spent his naptime awake, pleasantly sticking it to The Man together. One afternoon he pointed to the clock on the bedroom wall; its ticking was keeping him awake and he wanted a closer look at it. I took the clock down, brought it over, and showed him the plastic box on the back that held the battery and the mechanism. Then I turned it over and together, mystified, we watched the second hand go around.

I work out of an old building at the foot of a hill by the Hudson River. The building, a former brewery, houses an eclectic mix of local businesses, including a contracting company, a piano repairer, a children's dance studio, and various artists and musicians. The walls are thin, the floors are linoleum, and the whole structure is in the process of decay. At night I cover my computer with plastic sheeting in case the ceiling springs a leak or drops crumbs of grit. One morning I noticed that a mud wasp had begun building a nest on my ceiling. Another day, through the wall, I heard the owner of the business next door scold an employee, who happened to be his mother: "If I'm on deadline and I need more time, the first thing I do is figure out how much more time I need!"

Out in front, by the parking lot, is a man-made pond with a bench where I go sometimes to think. The pond is small, maybe a hundred feet across, with a concrete lip around the edge. Water—suburban runoff—enters at the far end from a gully thick with weeds and exits at the near end down a drainpipe. In early spring the water is clear enough that I can see the resident goldfish near the bottom, four feet down. By mid-May the surface has acquired a green film and by late June the pond is clotted with scum, offering little to see but much to consider.

The scum of the earth hardly get the credit they deserve. What we call scum are often cyanobacteria, formerly known as blue-green algae, a broad group of single-celled prokaryotes—organisms lacking a cell nucleus—that live in the water and thrive on sunlight. Cyanobacteria aren't the customary bacteria (your household germs don't photosynthesize) nor are they exactly algae (which are single-celled eukaryotes, with a nucleus). But they are everywhere; they make up a decent fraction of Earth's biomass and are the foundation of the food chain. Cyanobacteria are among the oldest forms of life on our 4.5-billion-year-old planet. They appeared at least 2.8 billion years ago, and perhaps as long ago as 3.8 billion years, before Earth's atmosphere contained oxygen;

in fact they are credited with single-cellularly creating the oxygen as a by-product of their photosynthesis. Somehow, at some point, the dry essence of space-time was internalized—embodied—by life. If the history of living time starts anywhere, cyanobacteria are as good a place as any to begin.

Having an endogenous clock is a useful adaptation. For one thing, the clock is an essential backup. In theory, an organism could do without one and instead fulfill all its timing needs, such as keeping its internal environment organized, by tapping directly and continuously into the twenty-four-hour rhythm of daylight—except that the organism would go adrift every night and on cloudy days. (Imagine if your radio-controlled clock lost radio reception after sundown and had no means of keeping time on its own.) Still, until the late nineteen-eighties most biologists assumed that microbes such as cyanobacteria don't have a circadian clock, for the simple reason that the average microbe doesn't live long enough to need one. A typical cyanobacterium divides into two new ones every few hours—more quickly and vigorously when the sun is shining, less so in the dark. Over a twenty-four-hour period, a parent cell can give rise to six or more subsequent generations, resulting in scores of cells. As Carl Johnson, a microbiologist at Vanderbilt University, put it to me, "What's the point of having a clock if you aren't going to be the same person the next day?"

For more than two decades Johnson has been at the forefront of research demonstrating that in fact bacteria do have a circadian timepiece, one that is astonishingly accurate. Moreover, the bacterial clock bears so little resemblance to the clock in the cells of animals, plants, and fungi that it begs the question of why the circadian clock evolved in the first place—and how the subsequent varieties of clock might be related.

Cyanobacteria create oxygen in the course of photosynthesis; many species also fix nitrogen, pulling it from the air and converting it into compounds that can be utilized by plants. Doing both at once is a challenge, as the presence of oxygen stifles the enzyme involved in capturing nitrogen. More complex, filamentous cyanobacteria can carry out these activities simultaneously by dividing the labor among its cells. But single-celled cyanobacteria don't have internal compartments, so instead they compartmentalize across time: they conduct photosynthesis by day and fix nitrogen by night.

The existence of that daily rhythm was one hint that microbes possess some sort of circadian clock. With several colleagues, Johnson deciphered the clock's mechanics, mainly through the study of *Synechococcus elongatus*, a species of cyanobacteria commonly used in lab experiments. Its clock is widely shared among diverse cyanobacteria, and similar aspects of the clock are seen in other microbes, but it bears little resemblance to what's found in higher organisms. At its core are three proteins, nicknamed KaiA, KaiB, and KaiC, after the Japanese Kanji character *kaiten*, which refers to the cyclical turning of the heavens. The key protein, KaiC, looks a bit like two doughnuts stacked one on top of the other or, aptly, like a cog in a clock. Occasionally KaiC interacts with one of the other two proteins, which alters its shape slightly and permits it to grab or release a phosphate ion. Eventually all three proteins converge to form a single, ephemeral molecule called a periodosome. Susan Golden, a microbiologist at the University of California at San Diego, refers to the interaction as "a group hug," and the embrace takes about twenty-four hours to achieve.

"It's almost like the gears of a clock moving around," Golden told me. There are several remarkable aspects of this setup, but the main surprise is its independence. In higher organisms, the circadian clock is driven by the rhythmic expression of the DNA; key genes in the nucleus prompt the construction of proteins in the cytoplasm that then turn off those genes in the nucleus. Cyanobacteria have no nucleus; their clock is a conversation among the proteins alone. Those proteins are manufactured by specific genes (knock out the genes and the clock eventually fails for lack of components) but the rate at which the protein clock ticks isn't tied to the rate at which the genes are expressed. In fact the tick rate of the protein clock is sufficiently independent of the cell's DNA that when the key proteins are removed from the cell and isolated in a test tube, they continue to enact their twenty-four-hour hug for days on end.

"In plants, animals, and fungi, the clock is a very vague thing," Golden said. "It's a summation of events, and there are lots of players milling around. What's extraordinary about the cyanobacterial clock is that it's a thing: it's a device. You can isolate it in a test tube and get it to do its thing."

Certain components of cells—the energy-generating mitochondria, for instance, and chloroplasts, where photosynthesis occurs—are thought to have once been free-swimming prokaryotes that were ingested but were never metabolized; they are internal symbionts, basically. I wondered if the protein clock had a similar history—whether it had existed, or perhaps still exists, independently in nature and was internalized by cyanobacteria, like a borrowed watch. Golden said no; scientists managed to replicate the clock outside the living cell only by using meticulous techniques in the lab. But its existence speaks to the durability and simplicity of the machinery, she said. Given the right container and just a handful of pieces, it doesn't take much for natural selection to produce an accurate watch—a watch, moreover, that is readily passed from generation to generation.

Indeed, when a cyanobacterium divides, the clock splits into two and continues ticking, without missing a beat. Two bacteria become four, eight, sixteen, millions—all identical, all containing identical clocks keeping the same original time, synchronized en masse. The clock consists of a heap of proteins interacting in a sack, the cellular membrane. When the sack divides, the proteins are divvied up too, so the mechanism remains intact, keeping the old beat in two new containers. Because the mechanism runs independently of the organism's DNA, the clock transcends the lifespan of any individual cell. One would hardly know from just looking, but the filmy surface of a pond, made up of billions of cyanobacterial cells, presents the unified face of a watch.

Some version of this clock has been found in several dozen other species of cyanobacteria. "There could be other organisms with other kinds of clocks," Golden said. "We don't know how many clocks are out there." With so many varieties of clock in service—in animals, plants, fungi, bacteria—biologists have come to wonder how deeply these clocks may be related. Two lines of thought have emerged. The first, what one might call the Many Clocks school, argues that because the twenty-four-hour rhythm of sunlight has been such a pervasive force of natural selection, and because the circadian clock is such a critical adaptation, countless forms of the circadian timepiece have evolved. "Different organisms had different things in their kitchens they could cook up a clock with," Golden said. "If it works, it works."

The other branch of thought, the Single Clock school, turns that argument on its head: the rhythm of sunlight has been such a pervasive selective force that once the original circadian clock evolved, it stayed evolved. This argument is harder to make; the disparities between clock types—between humans and plants, or plants and fungi, or fungi and cyanobacteria—seem too great to be easily reconciled. But Carl Johnson, for one, contends that they can be, eventually. He suspects that, hidden beneath the dialogue that comprises the clock of multicellular organisms—the transcription of genes and the translation into proteins—something like the protein clock of cyanobacteria ticks away, driving the apparent conversation. "I've been pushing the idea that transcription-translation may not be the core model," he told me. "Maybe the cyanobacteria are leading us to a new way of thinking."

Stare long enough at a pond of scum clocks and questions start to bubble up. For instance: Did the circadian clock evolve many times or only once? Why did it even begin? There are no provable answers, of course; natural selection covers its tracks. Still, almost certainly sunlight played a role in the clock's emergence. That the rhythm of the circadian clock and the length of the solar day align so closely, and so consistently and widely across the kingdoms of life, can't be mere coincidence.

Put yourself in a microbe's shoes and imagine what you could do with your own twenty-four-hour clock. It's a handy backup for when the sun fails you, but it's also an anticipatory device, almost an alarm clock; it provides a good estimate of when the sun will appear tomorrow, enabling you to prepare yourself for it. If you're photosynthetic, such a clock might allow you to ready your energy-harvesting machinery and perhaps gain an edge on other photosynthesizers—and so reproduce more successfully and bequeath your clock to future generations. This advantage might be less helpful near the equator, where the length of day and night are equal and the hour of sunrise and sunset never varies. As one moves north or south toward Earth's poles, however, the ratio of daylight to dark changes each day as the year advances, and a circadian clock would help anticipate the variation. Perhaps the clock enabled early organisms to expand their range, much as, in the seventeenth century, the invention of longitude and the mechanical clock helped the British explore the world's seas and colonize distant lands.

But as a selective force sunlight cuts both ways; it's something to be avoided as well as utilized. Ultraviolet radiation can do grave damage to a cell's DNA; the genome is most vulnerable during cell division, when the DNA unzips to duplicate itself. Conditions would have been especially perilous four billion or so years ago, before the planet developed the protective layer of ozone that now shields life from the sun's most hazardous rays. And cyanobacteria, which are largely credited with creating the planet's oxygen and its ozone layer—a feat that took at least a billion years—would have been most at risk. Lacking flagella, they can't move, so they would have been unable to sink into the shade of the water column. How did they reproduce without exposing their tender parts to ultraviolet rays?

A circadian clock might help; with it a microbe could arrange for cell division to occur at less hazardous times of day. Biologists call this the "escape from light" hypothesis. Although cyanobacteria seem to divide nonstop in the light—they run on solar energy, after all—they may place some temporal constraints on their reproduction. A study of three microbial communities in the wild—two were composed of algae, the third of a species of cyanobacterium—found that they photosynthesized all day but shut down the production of new DNA for three to six hours in the middle of the day, then started up again before sunset. The regions of them that are most susceptible to ultraviolet radiation effectively took a midday nap in the shade.

Modern plant and animal cells may contain a relic of this evolutionary history, in the form of specialized proteins called cryptochromes. These proteins are sensitive to blue and ultraviolet light; they're also part of the circadian clock in these organisms, helping to keep them synchronized with the natural cycle of daylight. The proteins are re-markably similar in structure to an enzyme called DNA photolyase, which uses the energy of blue light to repair DNA that has been dam-aged by ultraviolet radiation. Some biologists think that the enzyme's role may have evolved over time. Perhaps what began as a tool for fixing ultraviolet damage was co-opted and incorporated into the circadian clock, where, as a cryptochrome, it now serves a more managerial role, helping the organism sidestep solar damage altogether. The medic be-came a mediator.

If the escape theorists are right, then, the circadian clock was the world's first prophylaxis, the precursor of safe sex. Those organisms that could anticipate and avoid reproducing during the most hazard-ous hours of sunlight were rewarded with another generation, while the ill-timed and improperly equipped were genetically fried. Life or death—straightforward, Malthusian. When I look at my office pond, I don't immediately see a clockworks, but I suppose that's what it is. They may be scum, but we owe them the time of day.

February 14, 1972: Michel Siffre begins his second major experiment in time isolation, the longest in history. In Midnight Cave, near Del Rio, Texas, and with financing from NASA, he has established a subterranean laboratory of himself. A wooden platform supports a large nylon tent equipped with a bed, table, chair, various scientific devices, freezers of food, and seven hundred and eighty one-gallon jugs of water. No calendars, no clocks. He smiles for the news cameras, kisses his new bride, hugs his mother, then descends the hundred-foot vertical shaft that leads to his isolation. If all goes well, he will remain there for more than six months, until September. "The darkness is absolute, the silence total," he later writes.

Siffre counts his days by cycles, from waking hour to waking hour. The mornings are busy: on rising, he calls his research team up above and they turn on the lights he's had installed in the cave. He notes his blood pressure, rides three miles on a stationary bike, and shoots five rounds of target practice with a pellet gun. He attaches electrodes to his chest, to measure his cardiac rhythms, and to his head, to record the nature of his sleep; with a rectal probe he measures his body temperature. When he shaves, he saves his whiskers to be studied later for any hormonal changes. And he sweeps. The rocks around him are decomposing into a dust that settles everywhere; the dust is mixed with the guano of a former bat colony, so as it billows he tries not to inhale.

Siffre is interested to learn what happens to a person's natural body rhythms when isolated from time for so long. Studies by Jürgen Aschoff and other researchers indicated that some subjects, isolated for as long as a month, begin to inhabit a forty-eight-hour day, sleeping and waking for twice as long as a normal person. Might the crews of spaceships or nuclear submarines achieve and benefit from such a regimen? But all this measuring, the attachment and removal of probes and electrodes, the sifting of whiskers, quickly becomes tedious for Siffre. Before the first month is over, his record player, his principal source of diversion, breaks. "Now I have only books," he writes in his notes. Mildew is spreading, growing even on the dials of his scientific equipment.

The results of the tests and measurements will later show that for the first five weeks underground, Siffre lived on a twenty-six-hour circadian cycle. His body temperature rose and fell once every twenty-six hours and, although he was unaware of it, he slept and woke on that schedule too, rising two hours later each day and sleeping one-third of the time. As he did in the Scarasson, he became free-running: he lived the Rousseauian ideal, on a strictly endogenous timetable that engaged neither sunlight nor society.

On his thirty-seventh day underground, which by Siffre's reckoning is day thirty, something unprecedented happens. Without his knowing, his temperature and sleep cycle, which are already unglued from the solar day, become unglued from each other: Siffre stays awake long past his usual bedtime and then sleeps for fifteen hours, twice his usual sleep period. After that his schedule flips back and forth; sometimes he sleeps on a twenty-six-hour cycle, other times it lasts forty to fifty hours. All the while, his temperature cycle maintains a twenty-six-hour beat. Siffre is oblivious to it all.

Scientists have since learned that our sleep habits are only partly governed by the circadian cycle. Over the course of the day, the neurochemical adenosine accumulates in the body, inducing sleep; its buildup is called homeostatic pressure. You can override that sensation by taking a nap, which burns off some of the adenosine and pushes the sleepy sensation later into the night, or by toughing it out, perhaps by drinking caffeine, and trying to stay awake as long as possible. Once you're asleep, however, your circadian cycle takes over. In the early stage of sleep you fall into a deep slumber but as the night progresses you begin to dream. Dreaming, or rapid eye-movement sleep, is most likely to occur when the body temperature is lowest. For most people, this occurs a couple of hours before waking for the day. Thus, because body temperature runs on a circadian timetable, one is quite likely to wake up from a long dream before dawn, and to do so at close to the same time every day—at 4:27 a.m., for instance.

Put another way, adenosine puts you to sleep, if you let it, and the intensity of your sleep is determined by how long you were awake beforehand—how long you resisted the homeostatic pressure. But the predawn rise in your body temperature, which is a circadian matter, is what wakes you. You can manipulate the first factor to some extent, but not the second. How long you stay asleep depends on when you nodded off relative to the nadir of your body temperature. The closer to it that you fall asleep, the less you will sleep, even if you'd been awake longer than usual.

All this will be learned later, by scientists running isolation experiments in clean laboratories and on volunteers who are not nearly as sensory-deprived as Siffre. "I am living through the nadir of my life," he writes at one point. By day 77 his hands have lost the dexterity to string beads; his mind can barely string thoughts. His memory is failing. "I recall nothing from yesterday. Even events of this morning are lost. If I do not write things down immediately, I forget them." He panics severely when, after scraping the mildew from a magazine, he reads that the urine and saliva of bats can transmit rabies through the air. On day 79 Siffre picks up the phone. "*J'en marre!*" he shouts. I've had enough!

But he hasn't; his stay isn't even half over. He measures, monitors, probes, attaches and detaches electrodes, shaves, sweeps, bikes, shoots. Until one day he cannot. He disconnects himself from all cords and wires and thinks, "I am wasting my life in this stupid research!" Then he thinks of the valuable data that his colleagues may be missing while he is disconnected, and he plugs everything in again. He ponders suicide— he would make it look like an accident—then remembers all the bills for his experiments that his parents would be left to pay.

On the one hundred and sixtieth day Siffre hears the rustling of a mouse. During his first month in Midnight Cave, unnerved by the nocturnal shuffling of mice, Siffre had managed to trap and eliminate an entire colony of them. Now he is desperate for the company of just one. He names it Mus and spends days studying its habits and plotting its capture. Finally, on day 170, with a trap rigged from a casserole dish and baited with jam, he watches as his potential friend approaches warily. One more small step and . . . Siffre slams the dish down, his heart pounding with excitement. "For the first time since entering the cave, I feel a surge of joy," he writes. But something is wrong; he lifts the dish and sees that he has accidentally crushed the mouse. It dies as he watches. "The whimpers fade away. He is still. Desolation overwhelms me."

Nine days later, on August 10th, the phone rings: the experiment is over. Siffre will spend another month in the cave undergoing additional tests, but he can have human company at last. On September 5th, after more than two hundred days belowground, he returns to the surface, to a tumult of greetings and the scent of grass. Siffre has accumulated crates full of audiotape, miles of it, that await analysis. He has also developed weak eyesight, a chronic squint, and a half-million-dollar debt that will take him the next decade to pay off.

P ossibly the least useful thing one can bring to the Arctic in July is a flashlight. I brought two.

Even now I can't say why. North of the Arctic Circle, which begins at latitude sixty-six degrees north, a hundred and twenty-five miles north of Fairbanks, Alaska, the sun doesn't set from mid-May to mid-August. At its lowest it lingers just above the horizon and crawls around it, casting a pale light across mile after mile of undulating, boggy tundra, even at two o'clock in the morning. Summer is one long day. The entire ecosystem has evolved to take advantage of this season of constant daylight: to bloom, hatch out, feed, swim, mate, spawn, and hide again before the sun sets for the first time in late August and the available daylight begins to crash toward the weeks-long wintry night. I knew all this ahead of time. Yet somehow, somewhere, I envisaged a darkness I'd need to illuminate: in a cave I might explore; in the burrow of an Arctic ground squirrel I might peer into; under my cot in my dark, dark tent.

I'd come to the Arctic to spend time with biologists at the Toolik Field Station, at the edge of Toolik Lake, on Alaska's North Slope. The station, established in 1975, is a busy encampment of high-tech trailer labs and weather-hardened Quonset huts but otherwise is about as middle-of-nowhere as nowhere can be. To the south the Brooks Range is a jagged wall across the horizon. A hundred and thirty miles to the north is the town of Deadhorse, on Prudhoe Bay, at the coast of the Arctic Ocean and the northern end of the Trans-Alaska Pipeline; getting there entails a grinding, five-hour drive on the Dalton Highway, a wide gravel road dominated by tractor-trailers that toss up fist-sized rocks as they roar past.

In between are thousands of square miles of tundra and hundreds of shallow, teardrop lakes such as Toolik. Although tundra appears to be a bland and uniform landscape, it is a rich and varied ecosystem, a mix of mosses, lichens, and liverworts, sedges, grasses, and dwarf shrubs. A foot or two down is permafrost, but the unfrozen top layer is inhabited by voles, hares, foxes, ground squirrels, bumblebees, nesting birds, and other creatures. Every summer a hundred or so scientists and graduate students come to the station to probe the tundra, to gather specimens from lakes and streams, and to measure, weigh, and document. The landscape is not fragile as much as it is slow to change. Elsewhere the typical ecological study lasts no more than a few years, abridged by limited budgets and attention spans. Toolik represents a scientific commitment to learn how an environment functions across decades.

I was drawn to the science of days. In my application to stay at the station, I described my interest in circadian rhythms and the questions I hoped to explore by following the Toolik biologists around. "How does the light regime affect the metabolism and cycles of microbes and phytoplankton? How do these effects express themselves—via population distributions and growth rates, the availability of oxygen and nutrients, and other pathways—in the wider food web?" What I meant was, How are circadian rhythms manifest in the sparest of ecosystems, under the extreme conditions of a polar summer? What does biological time look like at its most bare?

Really, though, I just wanted to know what it felt like. In 1937 the explorer Richard Byrd spent four months, from April through July, alone in a shack in the frozen darkness of the Antarctic winter, taking meteorological readings. "This much should be understood from the beginning," he wrote in *Alone*, his memoir of that time. "That above everything else, and beyond the solid worth of weather and auroral observations in the hitherto unoccupied interior of Antarctica and my interest in these studies, I really wanted to go for the experience's sake. . . . I had no important purposes. There was nothing of that sort. Nothing whatever, except for one man's desire to know that kind of experience to the full, to be by himself for a while and to taste peace and quiet and solitude long enough to find out how good they really are."

I wanted to shed my clock. All the time-isolation experiments I'd read about involved being holed up in caves or in a dark, frozen shack. But two summer weeks of constant sunlight, in the wide-open air of

Alaska—that sounded appealing, just the sort of adventure that my quickly receding former self would appreciate. Never mind my two children, who would celebrate their second birthday in my absence; the sun was waiting to shine an eternal light on me.

Ten thousand years ago the most recent ice age ended and the last glaciers retreated from the North Slope; they left behind a sprawling network of streams and small, shallow, interconnected lakes, nearly all of them inaccessible by road. In 1973, biologists arrived—a small team from the Marine Biological Laboratory in Woods Hole, Massachusetts, who came to study the potential impacts of the pipeline, which was then under construction. They found Toolik Lake just off the gravel highway, near a pipeline-construction camp. They pitched their tents nearby and got to work, occasionally visiting the camp to do laundry or grab an ice-cream bar from the freezer. Eventually they relocated to the other side of the lake and set down roots; the station now spans several acres and has become the world's most advanced laboratory for the study of Arctic ecosystems.

One morning I followed John O'Brien, a freshwater biologist from the University of North Carolina, at Greensboro, to one of his field sites, a trio of small lakes a few miles due south of Toolik. It was an impossible distance on foot; tundra is a lumpy mix of spongy liverworts and hard tussocks of cotton grass, and walking across it for very far is exhausting, like tramping through a bog but with a much higher risk of turning an ankle. The station keeps a small helicopter for research outings, and O'Brien arranged for us to be deposited, along with three grad students, an inflatable dinghy, paddles, and backpacks full of sampling gear, on a slight rise above one of the lakes, which, at just a hundred yards wide, was barely a pond. When the helicopter had departed and the grasses had stilled, the mosquitoes closed in. The day, bright and breezeless, was unusually mild.

O'Brien had been part of the team that colonized Toolik in 1973 and he had returned almost every summer since, leaving his family behind for weeks to study the interactions between microscopic freshwater plants and the only slightly larger freshwater zooplankton that eat them. We typically consider ecosystems in terms of their living components—copepods, lichens, snow fleas, tiger moths, gray jays, Arctic graylings. But these life-forms are ephemeral bodies, temporary vessels for the perpetual flow of nutrients through them. Scientists come to Toolik with varying interests—botany, limnology, entomology—but ultimately they probe the same underlying biogeochemistry: carbon, nitrogen, oxygen, phosphorus, and other elements that cycle from soil to stream, leaf to air, rain to soil, and around again. These substances, meticulously measured through growth rates, respiration rates, and the weight of biomass, over time and across the landscape, provide a solid gauge of how the ecosystem as a whole is performing and changing.

After just my first day at Toolik it became clear that none of the researchers there were studying circadian biology, in the Arctic or anywhere. But more and more, all of them there were studying facets of a single concern, the undeniable warming of the planet's climate. The Arctic, with relatively few biological constituents, serves as a basic model for understanding how more complex ecosystems may respond to global warming. And the region is vital in its own right; at least ten percent of the world's terrestrial carbon is locked down in frozen tundra. As temperatures rise, how much of that carbon will be released? How much will be recaptured by plants to further their growth and how much will go into the atmosphere, to warm the earth still more? For the longest time Toolik was nowhere; increasingly it is at the heart of everything.

"In the early days, the back range was snowy on top and would be all summer long," O'Brien said. "This warm weather stinks." He was standing at the lake's edge, leaning on a dinghy paddle as if it were a staff and looking south toward the Brooks Range. O'Brien, sixty-six, was burly and inquisitive, with a shock of white hair and a bristling white beard. I found his presence reassuring, like an anchor, and I liked that he liked to tell stories, many of which began, "In the old days . . ." In the old days, you never saw a thunderstorm on the North Slope. In the old days, you would never think of wearing a T-shirt into the field, as he was doing now. In the old days, before laptops and G.P.S. locators and a dedicated machine-shop staff, you built everything at Toolik yourself.

"In the old days, you were so concerned with your animal needs, it brought out the animalness in you," he said. During his first summer in Alaska, he and several colleagues spent three months surveying the Noatak River Valley, another gem of pure nowhere in the state's vast reserve of it. They worked fourteen hours a day, seven days a week; the sun never set so neither did they. They soon grew sick of one another and all but stopped speaking. The cook cooked, minimally, but refused to wash dishes, so the group took to eating off an oilcloth, without plates or utensils. As an escape, O'Brien read *Sometimes a Great Notion*, Ken Kesey's novel about a logging family. The story and his surroundings so absorbed him that he began to think that he was a living character in the novel and always would be, and that his home life was the fiction. "We went totally bonkers," he said.

For two weeks at Toolik my home was a WeatherPort, a sleeping hut with plank floors and walls of ocher canvas. I slept on a mattress on a coil-spring bed frame and under a veil of mosquito netting. Like other station residents, I was encouraged to shower just twice a week, for two minutes, to conserve fresh water. There were amenities: high-speed wireless Internet access; a dining hall, open at all hours, that served polenta-crusted tilapia with banana-guava sauce; and, overlooking the glassy surface of Toolik Lake, a cedarwood sauna that was particularly busy after midnight.

But there was no darkness. For the first few days, I sprang from my sleeping bag to greet the day, made obvious by the sunlight illuminating the walls of my hut, only to check my watch and see that the time was 3:30 a.m. At night (or as I began to think of it, "night") I took to wearing eyeshades, as if I were on a transoceanic flight. When, according to my watch, true morning came, I would step outside and prod myself with the same irrelevant reminder: *Don't forget to turn off the light.*

Over evolutionary time, the ecological inhabitants of the polar regions have adapted to take this diurnal confusion in stride. In the Antarctic, chinstrap penguins tend to stick to well-worn routes as they shuffle from their colonies to the seacoast to dive and feed, and they are fairly rigid about their travel times; they depart on their trip at the start of their day with near twenty-four-hour punctuality, regardless of the temperature or the available light. (They return, at the end of the day, on a more relaxed timetable.) Bees in northern Finland, in the continuous daylight of summer, aren't active for the entire twenty-four hours. Their activity peaks at around noon and they lay off work by midnight, perhaps to warm their nests during the slightly cooler part of the day, or perhaps, by resting, to solidify their memories of that day's foraging effort. In these endeavors, at least, animals ignore the schedule of the sun and strictly obey the circadian clock they carry inside.

Arctic reindeer have adopted the opposite strategy. In 2010 Andrew Loudon, a researcher at the University of Manchester, in England, and his colleagues discovered that, in Arctic reindeer, two key clock genes don't oscillate in a circadian manner as they do in other animals. Most other complex organisms sleep, wake, and secrete hormones on a roughly twenty-four-hour schedule, and their circadian clocks are sufficiently sensitive to daylight that, even in the continuous wash of summer light, they are entrained to the physical day and kept in sync with it. Not the reindeer; the animal generates no internal circadian signals. Instead, it behaves in direct response to light: it rises when the sky brightens and sleeps when the light dims. It really has shed its clock—and is wholly a slave to the sun. "Evolution has come up with a means of switching off the cellular clockwork," Loudon has said. "There may still be a clock in there ticking away, but we have not been able to find it."

The biologists at Toolik have adopted a wide range of responses to the constant light. The season for collecting data is brief, so the researchers, with no darkness to slow them down, fan out across the landscape at all hours to gather, measure, synthesize, compare, and converse. On the Fourth of July I went to Deadhorse to see the Arctic Ocean; when I returned, at two-thirty in the morning, I found people in the dining hall eating lobster and filet mignon. Everyone at Toolik had a story about an insomniac they knew. One kept a camping mattress in his trailer lab so that he could sleep there at any odd hour; one summer, with the time he'd gained by sleeping less than usual, he built a foosball table and a sailboat. Another made a point of putting away his watch for the duration of his stay at Toolik and trying to ignore any reminder of the time; he ate and slept when he felt like it, and he worked incessantly. When he returns home at the end of the season, he confided to me, the night "kinda freaks me out." Another described a hike she recently had taken after dinner; she lost track of the time and when she returned to camp was surprised to find the kitchen staff setting up for breakfast.

Yet others obeyed the clock devoutly. "I need to go to bed when it's time to go to bed," one of O'Brien's grad students told me. We'd inflated the dinghy and were collecting water samples in the middle of the lake while O'Brien stood on the shore trawling for zooplankton with a small net. "If I waited until I was tired," the student said, "I'd be up all night, probably eating cake in the dining hall." O'Brien followed the clock closely and was famous for entraining his students to his schedule, or trying to. He expected them to show up in the dining hall every day for breakfast, but they devised a dodge: they'd stay up all night finishing their work, show up for breakfast, meet with O'Brien to update him on their progress and discuss their assignments for the day, then go back to bed. Twenty years went by before O'Brien found out about it.

If there was a common temporal landmark on that sea of light, it was breakfast. With few exceptions, everyone in camp scheduled their day around it; the meal began officially at six-thirty in the morning, and by six forty-five the mess hall was full. The draw was social as much as physiological: there were field plans to discuss, data runs to review, job openings to share, arguments to settle regarding who could most quickly inflate a Sevylor 66 dinghy. In theory one really could ignore all clocks at Toolik and live solely according to one's internal rhythm, or perhaps shed even that, but it was hardly practical. Any project involving more than one person required shared time to cement it: meet at the dock at noon; the chopper leaves for the Anaktuvuk site at nine o'clock sharp; salsa dancing in the mess tent at eight-thirty Friday night.

· · ·

When I look at my watch I see numbers, but with each passing day at Toolik the numbers mean less and less. Even the phrase "each passing day" has lost its significance. I simply inhabit one long day in which I nap occasionally and, on waking, am surprised to learn that, according to the numbers on my watch, I slept for several hours. Sleep has ceased to be the thing that divides one day from the next and has begun to feel optional. I find that I am spending more time in the station's phone closet, tethered to home by a T1 line.

Increasingly I dream about time. I dream that my children have broken my watch and scattered its shards on the floor. I dream that I'm walking across sand dunes when suddenly I slip to the bottom of a canyon and can't climb out; my friends don't know where I've gone and can't hear me shouting up to them. So I walk deeper into the canyon and under the monumental weight of the dunes, the daylight receding behind me, knowing that any moment the roof will shrug silently and bury me in sand.

The dream is almost certainly derived from my waking life, from a book I read in the library at home about a mountain climber who falls into a crevasse and breaks his leg. Unable to climb out, he crawls farther down into darkness and into the heart of the mountain. He sustains himself by licking moisture from moss until, miraculously, he finds an exit that leads out to the sunlit slope. But his campsite is miles away. So he keeps crawling: across an impossible labyrinth of natural ice bridges, down a boulder-strewn gulley, around the rocky shore of a lake. What propels him, he writes, is his watch. He lifts his head from the snow, picks out a landmark a hundred yards or two ahead, looks at his watch, and tells himself, "You have twenty minutes to get there," and crawls on. Except the voice he hears isn't his own—it's a disembodied one, some authoritarian Voice of All Time that echoes in his head and presses him forward. Near the end, shortly before his fellow climber finds him lying semiconscious close to camp, he lies awake at night under a field of stars, dehydrated, utterly disoriented, certain that he's been there for centuries.

The stillness in a place like Toolik is easy to mistake for timelessness. But time has always been here, in the scudding clouds, in the small motions of zooplankton, in the freezing and thawing, across eons, of the tundra. Change is coming more quickly now, and it is troubling. The average temperatures have been steadily rising at Toolik and across the Arctic. Thunderstorms, a rarity on the North Slope thirty years ago, are common. Scientists suspect that the retreat of sea ice on the Arctic Ocean is causing a shift in weather patterns that is making the region both drier and more prone to lightning. In 2007—the warmest on record at the station up to that point, and the driest in memory—lightning struck the tundra along the Anaktuvuk River, twenty miles from Toolik; it set a fire that burned for ten weeks and scorched nearly four hundred square miles, an area about the size of Cape Cod. It was the largest tundra fire in Alaskan and possibly global history. The summer I visited, Toolik researchers were busy trying to map the impact. With the insulating layer of peat now lost, more heat was entering the soil; in several places the underlying permafrost had partially melted, causing hillsides to slump and to leach soil and nutrients into the streams.

One morning I trailed Linda Deegan, an aquatic biologist from Woods Hole, as she waded into the Kuparuk River, which runs the length of the North Slope, from the Brooks Range to Prudhoe Bay. She'd been coming to Toolik since the nineteen-eighties to study the Arctic grayling, which migrates down the river in the spring and back up in late summer; it is the river's only fish and is central to the diet of some birds and larger lake trout along the way. In tracking them over the season and over the years, Deegan has tried to get a measure of how climate change might alter their numbers and the nature of their migration—when, how fast, how far—and the wider impact of those shifts.

Like many migratory animals, the grayling is genetically attuned to the sun. In the Arctic in the spring, each new day sees eight to ten additional minutes of daylight. The animal's circadian system registers the lengthening photoperiod, which triggers a cascade of physiological changes that prepare the fish for its journey downstream to breed. Deegan wonders about the insects that the grayling relies on to fuel its journey; their life cycle is keyed not to changes in sunlight but to water temperature. As annual temperatures rise, the insects may be hatching out slightly earlier in the season, perhaps before the grayling, locked to the intractable schedule of daylight, can arrive to fully benefit from them. Two life cycles, one driven by temperature and the other by light, are at risk of decoupling. Deegan hasn't quantified it, and the phenomenon hasn't been closely studied in the Arctic. "That's just my feeling," she said.

Scientists elsewhere are documenting a widening gap between the world of temperature and the world of time. In response to the warming spring, some migratory birds are reaching the Arctic and starting their breeding season as much as two weeks earlier than in past years, putting the latecomers at a new disadvantage; and the range of other birds is shifting northward to include the Arctic, where they compete with the local birds for resources. Some species are adaptable; many of the plants around Walden Pond now flower earlier and more abundantly than they did in Henry David Thoreau's time. But organisms whose seasonal behavior is more strongly driven by circadian cycles are more vulnerable. The pied flycatcher spends its winter in West Africa and flies in the spring to the forests of Europe to breed; its travel schedule, linked to the photoperiod, varies little. But the birds' young feed on caterpillars that are hatching out earlier in the spring than they did twenty years ago; by the time the flycatcher arrives in some areas there is little left for their young to eat, and their populations there have declined by ninety percent. It's as though the planet as a whole has begun to experience a kind of jet lag. Some species will make the transition to a warming climate and perhaps even thrive in it; they will migrate sooner, or later, or find other things to eat. Other species won't and that will be the end of them.

T imelessness, or something close to it, can be found in a deep cave or the far north, in the middle of the night or in unending daylight. But it's even easier than that to access: simply travel by plane, the farther the better.

Begin with the physics: you are several miles in the air, moving quickly and, bound by gravity, essentially falling. One of the peculiar consequences of Einstein's theory of special relativity is that time moves more slowly aboard a very swiftly moving object compared to the time of an observer who's standing still. Experiments have verified it: atomic clocks placed aboard jet planes have been found to tick more slowly—by a matter of nanoseconds over several hours—than a stationary clock on the ground. (In the plane itself, one second is still exactly one second long, identical in duration to the previous second; only observers in a nonmoving frame measure it as slower.) The effect is small but real. In March 2016, astronaut Mike Kelly returned to earth after spending five hundred and twenty days in orbit, circling the planet at nearly eighteen thousand miles an hour. In that time his earthbound twin brother, Mark, who was born first by six minutes, had aged by an additional five milliseconds.

Then there are the time zones: twenty-four in total, each an hour wide and spaced more or less evenly along Earth's lines of longitude, every fifteen degrees apart. Time zero is Greenwich, England, where the Royal Observatory is located. Because Earth is a rotating sphere, the sun can't illuminate all of it at once, so the daylight hours can't occur everywhere simultaneously; time zones are what make it possible for "twelve noon" to mean the same thing—the middle of the day, when the sun is about at its zenith—most everywhere in the world, even though it occurs in only one time zone at a time. Time zones came into use, piecemeal, in the nineteenth century, as a way to help railroads coordinate the schedules of their expanding rail networks. By 1929 most of the world had signed on to the hourly time-zone scheme, although some countries today have their time zones set on the half-hour and even, in Nepal, at the forty-five-minute mark. In 1949 the geographically sprawling China adopted the opposite strategy and reduced its five time zones to a single large one.

Nowadays, with air travel, we cross time zones regularly. In the seven hours it takes to fly from Paris to New York, one can erase the six-hour difference between the cities. Clocks are ultimately local; what time it is depends on where you are. If you're on a plane—moving at some fast speed, staring down at an unending canvas of ocean—your where and when are changing with every moment. My watch may still be set to Paris time, some hours behind me, while the informational map on the headrest in front of me reports the time in New York, which is still hours away. I'm in between for an indefinite—seemingly eternal— period of time.

There is a central time on our flight, up in the cockpit with the suprachiasmatic captain. The universally coordinated time of the world's many atomic clocks, sifted and weighted according to the advisory algorithms of the B.I.P.M. in Paris, is continuously transmitted via satellite into the guidance systems of moving cargo ships, rental cars, and planes. Out in the main cabin, however, it's every clock for itself. Some passengers doze, others eat. Some aim themselves toward the late-afternoon meeting that awaits them; others are recovering from their early-morning effort to catch the flight. And others still are lost in onboard-movie time, far away and ending happily. Traveling west, bathed in constant daylight, devoid of meaningful time cues, we follow our own decentralized hours.

How the brain's suprachiasmatic nucleus disseminates its time throughout the human body is still poorly understood. But the process takes time—hours to days. If you're subjected to a sudden shift in your light regime and are forced to adjust to a new schedule—the sort of thing that happens when you cross a few time zones, or even for the day or two following the switch to or from daylight saving time—your peripheral clocks don't fall back in line all at once or at the same rate. Your body ceases to be a synchronized confederacy of clocks and instead becomes, temporarily, a conflagration of temporally autonomous states. That's the essence of jet lag. When my suprachiasmatic nucleus lands in New York, my liver may still be on Nova Scotia time and my pancreas may be somewhere over Iceland. For a few days, my digestive system will be out of whack, as my brain directs me to eat food at hours when my organs aren't fully aligned to metabolize it. (The body recovers at a rate of about one time zone per day.) The result is gastroenteritis, a common complaint of long-distance travelers and airline pilots. Jet lag is not in your head; it's an ailment of your entire, desynchronized body.

The scientific literature sometimes refers to your body's peripheral clocks as "slave" clocks beholden to the suprachiasmatic nucleus. But they can behave autonomously, and under the right circumstances they're capable of synchronizing their circadian rhythms not to the master clock and the natural cycle of daylight but to orders received from elsewhere. It turns out that food sends a particularly strong message to certain components of the body's clock. Several studies in the past decade have demonstrated that eating meals on a regular timetable can shift the phase of the liver's circadian clock, causing it to ignore the light-based timetable relayed from the brain and to perhaps even send a message of its own back upstream. Mealtime, not solar time, comes to define the liver's day. "If you feed a lab mouse in the middle of its sleep cycle, it will soon learn to wake up shortly beforehand," Chris Colwell, a leading circadian researcher at U.C.L.A., told me. "I tell my students, If the pizza guy starts delivering to your house every day at four a.m., I guarantee you'll start waking up at three-thirty."

One way to minimize jet lag, then, especially after a long flight, is to avoid eating the airline meals as they're handed out by the flight attendant. Their protocol requires that they feed you every couple of hours, typically on a schedule defined by the clock of the city you departed from. In transit, absent the normal light cues, the liver will drive the circadian clock, cementing you further to the time zone you're trying to leave behind. Better to set your watch immediately to the time zone of your destination and schedule your meals as if you'd already arrived. "The standard advice we give to people who are traveling," Colwell says, "is to expose yourself to light and mealtimes, as well as social interactions, as soon as you can." He also advocates eating breakfast. "If humans work in any way like lab mice do," he said, "breakfast is important to keeping those signals, so you don't go all haywire when those light signals aren't present."

Colwell's research suggests that regular exercise may also help drive the circadian system. In his lab he found that the suprachiasmatic nucleus generates stronger signals in mice that are allowed to exercise on a running wheel than it does in less-active mice; the effect was greatest in mice that were allowed to run only early in their waking day. The biggest beneficiaries were mice that lacked a particular clock protein; when they exercised late in their day, the suprachiasmatic nucleus showed an improved ability to send its organizing signals to the heart, liver, and other organs. Running more made their clocks run better. It's too early to know if scheduled exercise might help humans to the same extent. But the idea is tantalizing, Colwell said, because the quality of our master clock declines with age. "I'm barely fifty and I'm having trouble sleeping through the night," he said. "And I'm getting more tired during the day." Even timekeepers get old.

Jet lag, at least, is temporary. But humans are finding other, more lasting ways to defy the standard division of daylight and darkness, and the effects are worrisome. Millions of Americans do shift work: they drive through the night, work the late shift at the shipping center, or keep a crazy schedule at the hospital. Many suffer from what circadian biologists call social jet lag, with consequences that are more than merely inconvenient or uncomfortable. One of the key functions of the circadian clock is to supervise the body's metabolism—to ensure that we eat when we're hungry and that our cells receive the nutrients they need at the right time. But many researchers are finding that people who habitually work off-hour shifts are more likely to be obese, be diabetic, or suffer from heart disease. Mounting evidence suggests that there's a strong link between circadian misalignment—a sleep-wake cycle that's out of step with one's circadian clock—and metabolic disorder, a suite of conditions, including diabetes, that result when the body's system for digesting food falls out of step with the process of producing and storing energy.

Millions of dollars are spent studying what we should eat, but *when* we eat may be equally important. Mice that eat when they should be sleeping—that is, at the wrong time in their circadian cycle—gain more weight than mice that eat at normal hours, one recent study found. Although most studies on circadian misalignment have looked at rodents and nonhuman primates, medical researchers increasingly are turning their attention to human subjects. In one Harvard study, ten human volunteers were trained to live on a twenty-eight-hour day. By the fourth day their schedules had inverted: they were awake, and eating, in the middle of the night. Four days later their schedule had inverted back to normal. Within ten days—the length of the study—the subjects' blood pressures had skyrocketed, their blood sugar levels were above normal, and three volunteers were classified as prediabetic. The confirmed cause wasn't lack of sleep; rather it was the fact that the subjects were consistently eating at times of day when their organs and adipose cells weren't primed to metabolize the food. "Even after just a few days, they showed striking changes in glucose metabolism," one of the study's authors noted. "The rapid onset within just a few days shows that such changes may even temporarily affect the millions of people experiencing jet lag every year."

The current obesity epidemic has many causes, including our sedentary lifestyle and less-than-exemplary diet. But the circadian research suggests another, less visible culprit: increasingly we are trying to colonize the wrong part of the day. "We have a perfectly good endogenous timing system that works based on the old rules," Colwell said. "It's crazy to think that just because we've invented electric lights we can ignore it."

S ooner or later, if the scientists are right, humans will go to Mars. This will be an undertaking. The planet is thirty-six million miles away and simply getting there, using present-day propulsion technology, will take six months. That's six months in a tin can with your companions and artificial lighting. There will be no windows, in order to minimize your exposure to cosmic radiation while traveling for so long beyond Earth's protective magnetic shield. (Not that you'd see anything out there anyway, except blackness and stars.) Researchers are already thinking about how to make such a trip tolerable to the passengers—which foods are healthiest and most flavorful, which activities will best stave off boredom, what to do in the event of a medical emergency. Still, we'll get there. You'll step out of the tin can and into the glare of the Martian summer sun, and then scurry into whatever habitation has been built—another windowless container with its artificial lights dimmed to conserve energy.

Our first day on Mars will be the longest one that humanity has ever known. The planet doesn't rotate on its axis quite as fast as Earth does, so every Mars day is 24.65 Earth hours long—that is, thirty-nine minutes longer than a day on Earth. That may not seem like much but it's thirty-nine more than the human circadian system was naturally selected to accommodate, and the new Martians would soon feel the ill effects. "It would be like traveling two time zones every three days," Laura Barger, a physiologist at Harvard Medical School and Brigham and Women's Hospital in Boston, told me. With Charles Czeisler, the director of the sleep medicine divisions at Harvard and at Brigham and Women's, and other colleagues, Barger has studied the circadian rhythms of orbiting astronauts and of the mission controllers who must maintain contact with them at all hours. In one study, they had volunteers try to adapt to a 24.65-hour-long day. "Their circadian rhythms aren't able to adjust,"

Barger said. "They have problems sleeping and everyone walks around with a bloodless pallor."

In 2007 Czeisler ran an experiment to see whether, by using artificial lights of certain wavelengths and at certain times of day, they could force the circadian clock to shift to a twenty-five-hour cycle, which would be more amenable to a life on Mars. A dozen volunteers spent sixty-five days in dimly lit rooms without clocks, windows, or other time cues. For the first three days they lived a twenty-four-hour day; then scientists tacked on an extra hour of light, effectively adding an hour of wakefulness to each participants' day. To help them adjust, the researchers turned up the lights toward the end of each extra-long day to a brightness roughly equivalent to sunset or sunrise and delivered two forty-five minute doses of it, an hour apart. After thirty days, the volunteers had successfully adopted and adapted to a twenty-five hour day.

Science demonstrated that it can be done, that the solar system— and its grip on human biology—can be temporally conquered, at least a little bit. What will tomorrow's humankind do with its extra hour? Work, probably. The researchers, in their paper, note that productive activities might include "caring for crops in a brightly lit greenhouse module." Afterward we'll have a drink, take in the windowless view, and click through our old photos of Earth.

On November 30th, 1999, thirty-seven years after his descent into the Scarasson cave, Michel Siffre begins his third and perhaps final time-isolation experiment. Now sixty years old, he aims to study how his circadian rhythms may have been affected by his aging. Again he chooses a natural cave: the Grotte de Clamouse, a limestone network in the southern Languedoc region of France. A wide wooden platform is again built in one particularly large cavern, on which another nylon canopy is erected. At the cave entrance, researchers, well-wishers, and the media cheer as Siffre, wearing a miner's helmet and lamp, removes his watch and, with a last turn and wave good-bye, tromps into the darkness.

His living quarters are bathed in the light of halogen lamps. From video footage that he shot himself, we see Siffre at a wooden workbench, eating salmon from a can and entering his mealtimes into a computer. His entries, activities, and well-being are monitored from a research room beyond the cave. Siffre wears green rubber boots and a red fleece vest, even when working out on a stair-climbing machine. He saves his urine in glass vials. He sleeps in a sleeping bag lashed to a lawn chair; when he wants, he can recline comfortably and read books off a crowded shelf nearby. He never talks to himself, but sometimes he sings.

On Monday, February 14th, 2000, Siffre emerges from his geologic womb. Cheers, applause, the flash of cameras. Once again he has demonstrated that, isolated from daylight, the human biological clock will free-run at a pace slower than Earth's rotation. Seventy-six days have passed since he went underground, but he thinks that only sixty-seven have transpired, that today is the fifth of February. In the early hours of January 1st, while the rest of the world greeted a new millennium (and sighed in relief that their computers had not come to a crashing halt), Siffre did nothing; by his count the date was December 27th. His New Year fell on the wider world's January 4th.

Years later, Siffre tells an interviewer that to be isolated for so long underground is to inhabit a seemingly eternal present: "It's like one long day. The only things that change are when you wake up and when you go to bed. Besides that, it's entirely black." Emerging from Clamouse, he confides to a reporter, "I've the impression that my memory has been impaired. I can't even remember what I did down there yesterday or the day before yesterday."

He exits into the sunlight. He is relieved to be here, out in the open, in a now that begins and ends. He says, "It's brilliant to see the blue sky again."

THE PRESENT

In hashish-intoxication there is a curious increase in the apparent time-perspective. We utter a sentence, and ere the end is reached the beginning seems already to date from indefinitely long ago. We enter a short street, and it is as if we should never get to the end of it.

—William James, *The Principles of Psychology*

How Long Is Now?

As I write, I am seated in the café car of a train, headed home after visiting a friend in another city. I occupy the booth at the forward end of the car and sit with my back to the front wall, facing the rear of the train. The whole car opens before me like a stage: at a nearby table two college students drink coffee and talk over their textbooks; at another table the conductor chats with the café attendant, who is on her break; at the far end several passengers huddle around a young man's laptop watching the tense final minutes of a football game. My eyes drift to the long panel of windows that runs the length of the car; in the deepening dusk I can still make out the silhouettes of houses and an occasional streetlamp as we go by. They appear abruptly at the edge of the window to my immediate right, streak down the length of the car, then disappear from sight and mind, followed by more streetlamps and silhouettes in a continuous, darkening stream. I entertain the thought that each passing light and house is entering into existence only just now, from some point just behind my right shoulder. They seem to emanate from me as I hurtle backward into the future, gazing through the present and into memory.

Lying in bed at home, in the dark hours before dawn, I have the opposite experience. The bedside clock ticks and one by one the sounds of the seconds take shape in the darkness in front of me like mile markers on a road at night. They approach, pass through me, then dissipate somewhere behind my pillow, leaving me to wonder where they come from and how exactly each one cedes to the next. "If you could choose an hour of wakefulness out of the whole night, it would be this," Nathaniel Hawthorne wrote. "You have found an intermediate space, where the business of life does not intrude, where the passing moment lingers, and becomes truly the present." I can't say where the road leads, but at this hour, and at this hour only, it feels as if I have all the time in the world to consider it.

For more than two thousand years the world's great minds have argued about the true essence of time. Is it finite or infinite? Is it continuous or discrete? Does it flow like a river or is it granular, proceeding in small bits like sand trickling through an hourglass? And most immediately, what is the present? Is *now* an indivisible instant, a line of pure vapor between the past and the future? Or is it an instant that can be measured—and if so, how long is it? And what lies in between the instants? How does one give way to the next: how does *now* become *next* or *later* or simply *not now*? "The instant, this strange nature, is something inserted between motion and rest, and it is no time at all," Plato remarked in the fourth century B.C.E. "But into it and from it what is moved changes to being at rest, and what is at rest to being moved."

A century before Plato, Zeno of Elea whipped these questions into a formidable set of paradoxes. Consider an arrow in flight. In any instant along its path, the arrow is at some fixed point; later it is at another fixed point. How—when, in what span of time—does it move from one to the next? To Zeno, an instant of time is irreducibly brief. The arrow can't move in such an instant; if it could—if it covered some fractional distance—then the instant must have some duration, a beginning and end. And if an instant has duration, it is divisible: in half an instant the arrow would move half as far, and so on down to indivisibility. Pity the swift Achilles, unable to ever reach his finish line. Aristotle, a student of Plato, was vexed by the paradox. "Motion is impossible," he wrote, summarizing Zeno's logic, "because an object in motion must reach the halfway point before it gets to the end." If motion is impossible, time is impossible too; it can't fly, because it never leaves the ground.

Aristotle tried to solve this conundrum with brute semantics, by arguing that time and motion are synonymous. Time isn't a firmament on which events play out; motion—the arcing sun, the arrow's flight—*is* time. He also argued that an instant has some actual, measurable duration in which motion unfolds: "Time is not composed of indivisible 'nows,' no more than is any other magnitude." But this opened a rabbit hole: does *now* do more than simply divide past and future? Is it always the same now, or does it change? And if it changes, when does that change occur? Surely not *in* now, Aristotle noted, for it "could not have perished in its own instant, since it was then."

Such infinitesimal concerns lead to existential caverns. If we can't explain how time advances from moment to moment, how do we account for change, novelty, creation? How does something emerge from nothing? How does anything—Creation, time itself—begin? The very self comes into question: how am I the same individual that I was a moment ago, or last week, or last year, or as a child? How do I change yet remain continuously me? In a comic Greek play that predates Zeno, one man approaches another to recover some money that he's owed. The debtor says, in effect, "Oh, but you didn't borrow from me! I'm no longer the same person I was then, any more than a pile of stones from which we've added and removed some pebbles is the same pile of stones." At this, the first man strikes the second in the face. "Why did you do that?" the second man asks, to which the first replies, "Who, me?"

If there's one thing that experts on time enjoy talking about nearly as much as time itself, it's how we talk. Time is encoded in the language we speak, as tense: past, present, future, and their various subcategories. We learn these early on, instinctively; by age two, children have mostly mastered the proper use of the past tense, although they may not consistently distinguish "tomorrow" from "yesterday" and "before" from "after." Pirahã, the language spoken by the Pirahã people of Brazil (and by a handful of linguistic scholars), contains few temporal references at all. Modern philosophers, for their part, divide themselves between tensers and detensers: those who contend that "past" and "future" are real qualities versus those who disagree.

To Augustine it was simpler still. Sooner or later almost any scientist who writes about the biology and perception of time quotes Augustine, because Augustine was really the first to talk about time as an internal experience—to ask what time is by exploring how it feels to inhabit it. Time may seem slippery and maddeningly abstract but it is also deeply intimate. Augustine proposed that time is in our every action, our every word; we need only stop to hear ourselves talk to grasp the urgency of the message. Indeed, the essence of time, all its texture and paradox, can be gleaned from a single line of speech, such as

Deus, creator omnium.

God, creator of all things. Say it aloud, or listen inside: in Latin, eight syllables, alternating short and long. "Each of these latter last twice as long as each of the former," Augustine wrote. "I have only to pronounce the line to report that this is the case." Yet how do we manage to make this measurement? The line is a series of syllables that the mind encounters in succession, one by one. How is the listener able to consider two syllables at once to compare their durations? "How am I to hold on to the short one, how am I to apply it to the long one as a measuring rod, when the long one does not begin to sound until the short one has ceased?" For that matter, how can one hold the long syllable in mind? Its duration can't be defined until it is completed—but by then both syllables are long gone. "Both have made their sound, and flown away, and passed by, and exist no more," Augustine wrote. "So what now exists for me to measure?"

In short, what is the present and where are we with regard to it? Not the present as in this century, this year, or even today, but the present right now before us, ever dissolving. If you have ever lain wake at night with a busy mind, listened closely to a burbling stream, or simply tried to tag your own thoughts as they enter and exit your awareness—that rivulet that William James called "the stream of consciousness"—you know what Augustine means. Borrowing from Aristotle, Augustine argued that the present is everything. Future and past don't exist: tomorrow's sunrise "as yet has no being"; his childhood is no longer. That leaves the present—an ephemeral duration, without extension, whose "only claim to be called 'time' is that it is slipping away into the past." Yet we clearly measure time. We can attest that the sound of one syllable lasts twice as long as another; we can judge the duration of a person's speech. When do we measure this time? Surely not in the past or future; we can't measure what doesn't exist. "We can only hope to measure it as it passes by"—that is, in the present. But how is that possible? How can one measure the duration of something—a sound or a silence—before it concludes?

From this paradox Augustine arrived at an insight so fundamental that the modern science of time perception takes it as a given: time is a property of the mind. When you ask yourself whether one passing syllable lasts longer or shorter than another, you're not measuring the syllables themselves (which no longer exist) but something in your memory, "something fixed and permanent there," Augustine noted. The syllables have passed but leave an impression that persists, that is still present. Indeed, he wrote, what we call three tenses are only one. Past, present, and future don't exist per se; they are *all* present in the mind— in our current memory of past events, in our current attention to the present, and in our current expectation of what's to come. "There are three tenses or times: the present of past things, the present of present things, and the present of future things."

Augustine plucked time from the realm of physics and placed it squarely in what we now call psychology. "In you, my mind, I measure time," he wrote. Our experience of time is not a cave shadow to some true and absolute thing; time *is* our perception. Words, sounds, events come and go, but their passage leaves an impression in the mind; time is there and nowhere else: "Either time is this impression, or what I measure is not time." Nowadays scientists explore this insight in laboratories with computer models, rodents, undergraduate volunteers, and multimillion-dollar magnetic resonance imaging machines. Augustine began where we all do, with an act of speech and a willingness to listen.

"Augustine isn't trying to come up with a philosophy or theology of time," my friend Tom tells me over lunch one afternoon. "He's trying to render a psychological account: what does it feel like to be in time?"

Tom is a friend from the neighborhood; our kids are about the same age and sometimes they play together and try to boss each other around. By day Tom is a theologian at a major university; by night he plays bass in loud bands and blogs about music, popular culture, and spirituality. I'm not sure I know what spirituality is, but Tom manages to make it sound both intellectually compelling and cool. We're in a restaurant in our small town and the place is otherwise empty. It's the Friday before Memorial Day and the weather outside is flawless, spring in full dilation.

Tom teaches Augustine as part of his introductory theology class, and his students, he says, identify with Augustine's intimate perspective. "We get trained into seeing time as something outside of us—time is what's ticking, what you see flashing," Tom says. "But it's in our heads, our souls, our spirit—our present." Time isn't merely observed; it is occupied, inhabited. Or perhaps it occupies us; at one point Augustine likens time to a volume—we are its vessel. So time isn't something to be discussed in the abstract; instead, look inside, listen to what you say, syllable by syllable, word by word. Come to know the container through the thing contained.

It's a subtle style of argument, one that later philosophers such as Heidegger will embrace as phenomenology: the study of conscious experience from the subjective point of view. "We can say this about ourselves because this is how we experience it," Tom says. It's a rhetorical device with an agenda, he adds: "It's meant to pull you in and change your perception, the way you step out in the world." Nowadays people attend weekend seminars in search of new ways to manage, experience, and relate to time. Augustine said, Attend to the words themselves.

Augustine's agenda is to effect a psychological change in the reader, to prompt you to transform self and soul, which you can do only by fully engaging the present. "You only taste the beautiful in the particular, in the transient, in the timely, in relation to what will pass away," Tom says. "The question is, How do you make of that taste a spiritual exercise? How do you do the right things in time?" We're accustomed to thinking of time as a quantity to spend or a tool to wield in the quest for self-improvement. For Augustine, time is something to reflect on and in; the spiritual imperative isn't to make the best use of time but to better inhabit it—to live within the syllable. "A lot of what I get from my students is what they get from their parents," Tom says. "They feel that they have to do the most they can in the time they have. It's a matter of durations, of potential parcels of productivity. One way to experience time is as the rungs of the ladder. A different way is, time is the thing you're climbing up into."

The restaurant closes at three; we're the last patrons, and the staff mill around with a nudging restlessness. Tom will head home to take over childcare for the afternoon. Together we're absorbing an early lesson of parenthood: children are terrible time managers; if it's not happening right now, by their clock, it's cause for a meltdown. I'm learning that part of my job is to introduce the vocabulary of *later* and *wait*. In fact it's becoming clear that a great deal of parenting will boil down to temporal education: teaching our kids to tell, keep, respect, cultivate, organize, and administer time but also to occasionally disregard all those mandates.

Spring is slipping toward summer and for the past several weeks Leo, now four years old, has been waking earlier and earlier. Maybe it's the light that wakes him; by 5:30 a.m. the pale of dawn is lightening the blinds and the robins have erupted into chorus. More likely it's his bladder, which goes off like clockwork. I hear him shuffle into the hall and into the bathroom, then back to his room. He won't stay there. While his brother sleeps in a bed just inches away from him, with stuffed animals crushed against his face to keep out the light, Leo, bright-eyed, tiptoes into our room and hovers by the bed. "I wanna go downstairs and play a game," he whispers. He means: *You come too.*

I've never been an early riser, and in winter especially I couldn't entertain the thought of going downstairs in the still-dark to a cold playroom. We tried technology. The boys couldn't yet tell time so we bought a clock that resembles a traffic light; you set it for a particular hour, say 6:45 a.m., at which time the light changes from red to green—the child's signal that it's okay to rise, make a racket, and start the day. Tom said it had worked for his daughter. But she doesn't have Leo's wiring or Leo's brother. The device had exactly the opposite effect from what we intended. Leo would rise early, as usual, go back to his bed, then lie awake transfixed by the red light and his own mounting impatience with it. Several times in the next hour he'd pad into our room to announce in a whisper, "It's not green yet." Or he'd flop around and sigh so loudly that he'd wake Joshua, and soon the two of them would be chattering and laughing at the mystery of the red light's intransigence. By the time 6:45 arrived and the light turned green, an event they greeted with a cheer, their official release came as a relief to everyone.

So I let Leo go quietly downstairs by himself, or sometimes Susan followed him down and I mashed the pillows over my face and fell back asleep. But increasingly I find myself slipping out of bed first and heading downstairs. The boys will finish preschool soon; in September they'll enter kindergarten at the public school and their lives will begin to bloom furiously outward, away from us. It's gradual, of course, and barely perceptible. Still, lately, it's hard not to feel as if we're on the brink of a transformation, and the days have taken on a crystalline quality, as if we've already begun to view them from memory. The boys sense it too, if only from us. A quiet half hour with one boy alone—by request, no less—is a gift. So Leo and I sit on the hardwood floor, with the back door open to the sound of the robins, and play another round of bingo, or checkers, or (so help me) Mousetrap, until Joshua comes shuffling in, gruff and rumple-headed—like me without coffee—and issues his own directives. These days, these hours, I see now, are sweet and few, to be slept through only by the young.

Villiam James can't sleep.

The year is 1876 and the young James, recently appointed to Harvard as an assistant professor of the nascent science of psychology, lies awake thinking about his future wife, Alice Gibbens, with whom he is deliriously in love. "Seven weeks of insomnia outweigh many scruples," he tells her when he finally pours forth his desire in a letter. A decade later, he frets in the dark about his years-long, two-volume, twelve-hundred-page book, *The Principles of Psychology*, which will become a classic almost as soon as it is published, in 1890. (According to Robert Richardson's biography of James, *In the Maelstrom of American Modernism*, James's insomnia worsened when his writing was going well; in the late eighteen-eighties he often resorted to using chloroform to fall asleep.) Perhaps, tossing, James wonders about the efficacy of the "mind cure" administered to him by Annetta Dresser, an adherent of what she calls the Quimby System of Mental Treatment of Diseases, after its originator, the late Phineas Quimby, a clockmaker who believed that physical ailments originate in the mind and could be alleviated by some combination of hypnotism, conversation, and right thinking. "I sit down beside her and presently drop asleep whilst she disentangles the snarls out of my mind," James tells his sister. Perhaps, awake in the dark, James wishes he had followed his physician's advice to try a bigger pillow.

Or perhaps he lies there absorbing the present. "Let any one try, I will not say to arrest, but to notice or attend to, the present moment of time. One of the most baffling experiences occurs. Where is it, this present? It has melted in our grasp, fled ere we could touch it, gone in the instant of becoming." His *Principles of Psychology* tackles a wide array of subjects, including memory, attention, emotions, instinct, imagination, habit, the consciousness of self, and "automaton theory," the persistent notion, of which James disapproved, that within our neural machinery lies some sort of homunculus, or mini-man, "that offers a living counterpart for every shading, however fine, of the history of its owner's mind," he wrote.

Among the more influential chapters is one on the perception of time; it's a deft synthesis of recent investigations by other researchers and of James's own thoughts on the subject. In Europe, scientific interest was shifting from pure physiology—the study of the body's mechanics—to the neurological signaling underneath, and from strict philosophy to a more rigorous study of mind and cognition. In 1879 the first laboratory for experimental psychology opened in Leipzig, Germany, under Wilhelm Wundt, who sought to quantify sensation and inner experience. "The exact description of consciousness is the sole aim of experimental psychology," Wundt wrote. Time perception was central to that study. James didn't believe in consciousness per se, by which he meant that it should not be addressed as some extramolecular "mind-stuff." Still, he felt, whatever exactly consciousness is, one could get a decent look at it by examining how we perceive time. James often described time through first-person experience because he considered it the best seat from which to accurately address the subject.

Sit quietly, he proposed. Close your eyes, turn off the world, and try to "attend exclusively to the passage of time, like one who wakes, as the poet says, 'to hear time flowing in the middle of the night, and all things moving to a day of doom.'" (James was quoting Tennyson.) What do we find there? Likely very little: an empty mind, a sameness of thought. If we notice anything, he says, it's a sense of the moments blooming one after another—"the pure series of durations budding, as it were, and growing beneath our indrawn gaze." Are we experiencing something real or is it an illusion? To James, the question speaks to the true nature of psychological time. If the experience is to be taken at face value—if one can truly grasp a blank moment as it emerges—then we must possess "a special sense for pure time." By this logic, pure time is empty, and an empty duration suffices to stimulate the senses. But suppose instead that one's experience of a budding moment is an illusion; in that case the impression that time is passing is a response to whatever is filling that time and to "our memory of its previous content, which we compare to the content now." The question is, is time anything without something in it? Is time a container or the things contained?

For James, time is in the contents. We can't perceive empty time any more than we can intuit a length or distance with nothing in it, he wrote. Look up into a clear blue sky: how far away is one hundred feet? How far is a mile? With no landmarks for reference, one can't say. It's the same with time. If we perceive time's passage, it's because we perceive change, and for us to perceive change, the time must be somehow filled; an empty duration alone won't stimulate our awareness. So what fills it?

Simply, us. "The change must be of some concrete sort—an outward or inward sensible series, or a process of attention or volition," James wrote in *Principles*. A seemingly empty moment is never truly so because, in stopping to consider it, we fill it with a stream of thoughts. Close your eyes, shut out the world, and still you see a film of light inside your eyelids, "a curdling play of obscurest luminosity." The mind fills in the time.

James is circling an idea raised centuries earlier by Augustine and, before that, Aristotle—that time is very much a property of the mind. James might not go so far as to say that time does not exist beyond one's perception of it, but he would emphasize that what the brain serves up is a *perception* of time, not time itself, and that that's as close as we'll get—there is no experience of time other than our subjective one. That may sound almost tautological but it's not far from where many contemporary psychologists and neuroscientists have landed. The average person is aware that time seems to speed up or slow down in certain situations, and it's easy to imagine that these impressions arise because somewhere in there, somehow, the brain is tracking how long a given stretch of time actually takes. But that clock may not exist. The brain may not time the real world, as computers do; it may only time its own processing of that world.

In any case, we can never quite escape ourselves. "We are always inwardly immersed in what Wundt has somewhere called the twilight of our general consciousness," James reflected. "Our heart-beats, our breathing, the pulses of our attention, the fragments of words or sentences that pass through our imagination, are what people this dim habitat. . . . In short, empty our minds as we may, some form of changing process remains for us to feel, and cannot be expelled."

Time is never empty, because we restlessly occupy it. Yet even that formulation gives time too much credit. I sit quietly, eyes shut, or lie awake in bed in the predawn hours, watching empty time flow. "We tell it off in pulses," James wrote. "We say 'now! now! now!' or we count 'more! more! more!' as we feel it bud." Time seems to flow in discrete units—it seems somehow independent and self-contained—not because we perceive discrete units of empty time, James wrote, but because our successive acts of perception are discrete. *Now* arises again and again only because we say "now!" again and again. The present moment, he contended, is "a synthetic datum," not experienced as much as manufactured. The present isn't something we stumble into and through; it's something we create for ourselves over and over again, moment by moment.

So much—all that matters, Augustine contends—unfolds in a sentence. Imagine reciting a poem or psalm by heart. As the words issue forth, the mind strains to recall what has been said and reaches forward to grab what remains to say. Memory is in the service of expectation: "The vital energy of what I am doing is in tension between the two." *The vital energy*—that's the essence of Augustine, and of you too, right now, as you absorb these words, strive to remember, and wonder what comes next. "Time is nothing other than tension," Augustine notes, "and I would be very surprised if it is not tension of consciousness itself." All these centuries later, scientists are still struggling to define consciousness, the self, and time. Augustine linked the three through language. You approach time only by trying to measure its passage as a sentence unfurls; there your mind is taut, present. And only in the present, in the act of attending, do you glimpse what you are. For Augustine, *now* is a spiritual experience.

James added a twist. He declared all three tenses—future, past, and present—nonexistent, and he invoked a fourth, what he called "the specious present." (He borrowed the term from E. R. Clay, which was the pseudonym for a retired cigar magnate and amateur philosopher named E. Robert Kelly.) The true present is a dimensionless speck; the specious present, in contrast, is "the short duration of which we are immediately and incessantly sensible." It is time enough to recognize a bird in flight or a shooting star, to assimilate all the notes of a bar of a song or the words in a spoken sentence. Never mind Zeno and his paradoxes or Kant's notion that we can somehow intuit the a priori nature of time; forget about past, present, and future. The only thing worth discussing about the present is our awareness of it, which effectively defines the specious present.

As I watch a bird in flight, read a line of a poem, or listen to the bedside clock at night, what can I say about this specious present? James noted that it is (or rather, appears in the awareness to be) constantly changing. "Any account of our perception of time must account for this aspect of our experience." Like Augustine, James noted that, to be aware of change, one must call on memory; to confidently say that a clock *is ticking* or a bird *is flying*, we hold in mind the awareness that the activity began or was under way a moment ago and now continues. A recognition of the present invokes some aspect of the immediate past, so this awareness must unfold over some brief period. "In short the practically cognized present is no knife-edge, but a saddle-back, with a certain breadth of its own on which we sit perched, and from which we look in two directions in time," James wrote. "The unit of composition of our perception of time is a duration, with a bow and a stern, as it were—a rearward- and forward-looking end. . . . We seem to feel the interval of time as a whole, with its two ends embedded in it."

The specious present, then, is a proxy measure of consciousness.

James offered a mix of metaphors: it is a boat, a gabled roof, something (a length of rope?) with "a vaguely vanishing backward and forward fringe"; it even "stands permanent like the rainbow on the waterfall." What matters is the flow of thought beneath it, the stream of consciousness. Your consciousness always contains several ideas or sense impressions at once. You don't experience event C, followed discretely by D, then E, and so on; instead you experience CDEFGH, with the first events eventually fading out of the present and new ones coming into play. The contents overlap; an awareness of some other part of the stream is always mixed in with the strict present. If, instead, consciousness were merely a series of images and sensations strung together like beads, we'd be unable to acquire knowledge or experience, and all we could know would be the present instant. James quotes John Stuart Mill: "Each of our successive states of consciousness, the moment it ceased, would be gone forever. Each of those momentary states would be our whole being." Our consciousness, James added, "would be like a glow-worm spark, illuminating the point it immediately covered, but leaving all beyond in total darkness."

James doubted that a practical life was possible under such circumstances, although it was "conceivable." It is more than that. In 1985 an accomplished conductor and musician named Clive Wearing suffered an attack of viral encephalitis that damaged several lobes of his brain, including his entire hippocampus, which is essential for recalling memories and laying down new ones. Subsequently he could walk, converse brightly, shave and dress himself, even play the piano. But he could remember very little and, three decades later, still can't: not his name or the names of people around him, nor which foods have which flavors, nor any thought prior to the sentence that he is in. By the time he supplies an answer, he's already forgotten the question. "A virus had caused holes in Clive's brain," his wife, Deborah, later wrote in the *Telegraph*. "His memories had fallen out." He doesn't remember her name, either, but he greets her ecstatically, with the embrace of the long lost, as he does no one else, even if she's only returning from a brief trip to the next room. He phones her anxiously at home, unaware that she was in his company just a few hours earlier. "Get here at dawn," he urges her. "Get here at the speed of light."

For Wearing, there is only the specious present. "He's stranded, if you like, on this tiny scrap of time," Deborah told the BBC, in one of the many documentaries and articles about him. Deborah has also written a book about her husband. One day, she writes, she found him studying a chocolate. He was holding it in the palm of one hand and, with the other, covering and uncovering it every few seconds, peering closely.

"Look," he said. "It's new!"

"It's the same chocolate," she told him.

"No . . . look! It's changed. It wasn't like that before . . ." He performed the trick again for himself. "Look! It's different again! How do they do it?"

Everything and everyone is perpetually new, including his own self, as if he were first waking to the world. "I can see you!" he exclaims to Deborah on another occasion. "I'm seeing everything properly now!" Or, "I've never seen anyone at all, I never heard a word until now, I've never had a dream even. Day and night the same: blank. Precisely like death." He says this again and again, in only slightly varying form, and has for years. "I haven't heard anything, seen anything, touched anything, smelled anything. It's like being dead." Unless his mind is otherwise engaged, that is his experience of life.

Yet the awareness of waking, of stepping into the present, is so momentous that Wearing has made a written note of it again and again. He writes the time—10:50 a.m—and records his insight: "Awake for the first time!" He notices a similar entry in the line above, made a few minutes earlier; he checks his watch, then crosses out the earlier entry as though it were written by an impostor and underlines the current one. Pages are filled with such entries, all but the current one struck through. The diary, if that is the correct term, now fills thousands of pages, dozens of volumes, each moment of awakening announcing its primacy over every one that came just before.

> 2:10 p.m.: *This time properly awake . . .*
> 2:14 p.m.: *This time finally awake . . .*

2:35 p.m.: This time completely awake . . .

9:40 p.m.: I awoke for the first time, despite my previous claims.

And I awoke properly at 8:47 a.m.

And completely at 8:49 a.m. And became aware of the problems of understanding me.

T he time traveler in *The Time Machine*, the novel by H. G. Wells, has a story for his dinner guests, of how he built a device that would spin him across time; and of how just moments earlier— before he encountered the effete Eloi and brutish Morlocks in the year 802,701, before he caught sight of an all-but-lifeless beach some thirty million years hence, before he entered the parlor just now asking for a drink—he had sat upon its saddle, pushed a lever, and "flung myself into futurity."

There is a feeling exactly like that one has upon a switchback—of a helpless headlong motion! . . . As I put on pace, night followed day like the flapping of a black wing. The dim suggestion of the labora-tory seemed presently to fall away from me, and I saw the sun hop-ping swiftly across the sky, leaping it every minute, and every minute marking a day. . . . The slowest snail that ever crawled dashed by too fast for me. . . . I saw trees growing and changing like puffs of vapor, now brown, now green; they grew, spread, shivered, and passed away. I saw huge buildings rise up faint and fair, and pass like dreams. The whole surface of the earth seemed changed—melting and flowing under my eyes. . . . My pace was over a year a minute; and minute by minute the white snow flashed across the world, and vanished, and was followed by the bright, brief green of spring.

Published in 1895, *The Time Machine* appeared at a moment of nov-elistic fascination with the notion of time travel. Most trips to the future or past happened unexpectedly, by uncertain means. The protagonists of *Looking Backward* and *News from Nowhere* fall asleep in the nine-teenth century and, after very long naps, wake up in the twenty-first. In *A Crystal Age*, the traveler wakes thousands of years after (he's fairly certain) falling from a cliff. In *The British Barbarians*, an anthropologist from the twenty-fifth century arrives somehow in Surrey wearing "a well-made grey tweed suit." *The Time Machine* is unusual in that its most compelling character is the mode of travel and, in a sense, time itself. The traveler isn't a passive agent; he aims himself at the moment of his choice. Nor does he simply arrive: he accelerates through every moment between now and then. In his hands, time is scaled and fungible; the specious present can be dilated to encompass seasons, human lifetimes, eons. The perceived present is no more than that—perceived. By altering perception, the traveler alters time.

Wells was firmly grounded in contemporary scientific theory. At university he studied biology under T. H. Huxley, and he clearly read *Principles of Psychology,* as nearly everyone in his circle did. In 1894, in the *Saturday Review,* he published a critique of contemporary psychology that displayed a solid grasp of the literature on memory, consciousness, visual perception, suggestion, and illusion. (One modern scholar, after dissecting the chronology in *The Time Machine,* makes a compelling case that the time traveler's dinnertime story is actually a hoax that he perpetrated on his guests—and that he dreamed up the story during a nap following a jaunt that afternoon on his three-wheeled bicycle.) The opening chapter of *The Time Machine* is effectively a short course on then-current notions of time perception. "There is no difference between Time and any of the three dimensions of Space except that our consciousness moves along it," the time traveler tells his guests, then launches into his theory of time as fourth-dimensional geometry, a theory that Wells is thought to have plucked from a lecture given, in 1893, at the New York Mathematical Society. "You cannot get away from the present moment," a guest objects at one point, to which the time traveler responds, "We are always getting away from the present moment." When the moment arrives to send a model of the time machine on its maiden voyage, it's the psychologist who flicks the switch.

William James kept assiduous note of everything he read, but there's no mention of *The Time Machine.* He read most everything else, from Augustine to *Tristram Shandy* to *Dr. Jekyll and Mr. Hyde.* ("The man is a magician," he wrote of Robert Louis Stevenson.) In his correspondence with Wells, James praised his *Utopia* and *First and Last Things* and compared him to Kipling and Tolstoy; Wells absorbed James's pragmatic philosophy and referred to him as "my friend and master." In 1899, by one account, they crossed paths at a party at Stephen Crane's house that involved late-night poker. Richardson's biography describes an occasion, several years later, when Wells stopped by to pick up James at the English home of his brother, Henry. Henry was agitated; he'd caught William standing on a ladder peeking over the garden wall and trying to catch a glimpse of the novelist G. K. Chesterton, who was staying at the inn next door. "It was most emphatically the sort of thing that isn't done," Wells recalled.

But it was the sort of thing that William often did. He was impulsive and would have headed for the ladder without a second thought, as if there was no time to waste. He went up stairs two or three at once. "He was a man in a hurry all the time," Richardson told me. He referred me to Henry James's autobiographical *A Small Boy and Others*, which was published in 1913, three years after William's death at the relatively young age of sixty-eight. Henry wrote that William "was always round the corner and out of sight," which Henry meant figuratively (William was a year older) but which also probably applied literally. "He was very alive all the time, right on the edge of it—and right on the edge of nervous collapse," Richardson said of William. "I don't think he felt he had a lot of time. And he didn't."

One evening in the late summer of 1860, the members of the Russian Entomological Society gathered in St. Petersburg for their first meeting. The keynote address would be delivered by the august German zoologist Karl Ernst von Baer, whom history mainly remembers as a curmudgeonly opponent of the Darwinian notion that all living organisms evolved from common ancestors. But Darwin himself greatly admired von Baer, who was a fierce intellect and a groundbreaking biologist and observer. Von Baer was first to argue that all mammals, including humans, arise from eggs, a conclusion he reached after tediously examining tiny, shapeless masses of embryonic chicks and other creatures through his microscope and marveling at the fact that vastly different organisms could arise from similar nascent forms.

The subject of his talk—"*Welche Auffassung der lebenden Natur ist die richtige? Und wie ist diese Auffassung auf die Entomologie anzuwenden,*" or "Which conception of living nature is the correct one? And how does it apply to entomology?"—might seem like an odd and elusive one for any crowd, much less a crowd of insect enthusiasts. But in the course of it von Baer discussed a matter that had circulated among philosophers since the seventeenth century and had recently entered the conversation of natural scientists: how long is now?

Nothing lasts, von Baer told his audience. What we mistake for persistence—the seeming permanence of mountains and seas—is an illusion derived from our short lifespan. Imagine for a moment "that the pace of life in man were to pass much faster or much slower, then we would soon discover that, for him, all the relations of nature would appear entirely differently." Suppose a human's lifetime, from birth to senility, lasted just twenty-nine days, one-thousandth its normal length. This *Monaten-Mensch*, or "man of the month," would never see the moon go through more than one full cycle; the concept of seasons and of snow and ice would be as abstract as the Ice Age is to us. The experience would be akin to that of many creatures, including some insects and mushrooms, that live for just a few days. Now suppose our lifespan were a thousand times shorter still and lasted just forty-two minutes. This *Minuten-Mensch*, or "man of minutes," would know nothing directly of night and day; flowers and trees would appear unchanging.

Consider the opposite scenario, von Baer went on. Imagine that our pulse, instead of speeding up, were to beat a thousand times slower than its normal rate. If we assume the same amount of sensory experience per beat, "then the life time of such a person would reach a 'ripe old age' at approximately 80,000 years. A year would seem like 8.75 hours. We would lose our ability to watch ice melt, to feel earthquakes, to watch trees sprout leaves, slowly bear fruit and then shed leaves." We would see mountain ranges rise and fall but overlook the lives of ladybugs. Flowers would be lost on us; only trees would make an impression. The sun might leave a tail in the sky like that of a comet or a cannonball.

Now multiply this life a thousand more times, to produce a man living 80 million years but having just 31.5 heartbeats and 189 perceptions in one Earth year. The sun would cease to appear as a discrete circle and would instead appear as a glowing solar elliptic, dimmer in winter. For ten pulse beats of the year Earth would be green, then white for ten more; snow would melt in a heartbeat and a half.

Through the seventeenth and eighteenth centuries, the increasing use of the telescope and microscope led to consideration of what might be called the relativity of scales. The cosmos was bigger than imagined, in both directions; it blossomed both out and in. The human perspective began to lose its sense of privilege: our outlook might be just one of many. Suppose, the philosopher Nicolas Malebranche posited, in 1678, that God had created a world so vast that a single tree would appear enormous to us yet seem normal to that realm's inhabitants—or, conversely, a world that appears tiny to us yet yawns in the eyes of its minuscule residents. "*Car rien n'est grand ni petit en soi,*" Malebranche wrote; nothing is big or small in and of itself. Jonathan Swift soon captured the idea in a novel; the outlook of the Lilliputians and that of the giant Brobdingnagians are equivalent in their detail and expanse.

So it is with time. "Imagine a world made up of as many parts as our own that was no bigger than a hazelnut," the French philosopher Étienne Bonnot de Condillac wrote in 1754. "It is beyond doubt that the stars would rise and set there thousands of times in one of our hours." Or imagine a world that dwarfs ours in its vastness: a lifespan in our world would seem but a flicker to the beings of that larger realm, while to residents of Planet Hazelnut our lives might last billions of years. The perception of duration is relative; a moment to one eye may be several to another.

To some degree this is wordplay. If we define a day as a single rotation of Earth on its axis, then one day always lasts exactly one day to human, mite, and hazelnut alike. (A circadian biologist would point out that in fact the day is genetically inscribed in each of us, hazelnut to human, whether we're conscious of it or not.) Condillac's point, however, was that to the Mites of Hazelnut, one day may not be a useful, or even a perceptible, span of time. That thought contains a notion of time that is still very much in play today: our estimate of how long a moment seems to last is shaped by the number of actions or ideas that pass through the mind as the moment unfolds. "We have no perception of duration but by considering the train of ideas that take their turns in our understandings," John Locke argued in 1690. If you experience many sensations in a brief period, then that duration, being densely filled, will feel longer while you're in it. An instant may seem dimensionless to us, Locke wrote, yet there might be other minds capable of perceiving it and we could have no more awareness of them than "a worm shut up in one drawer of a cabinet hath of the senses or understanding of a man." Our mind, moving only so quickly, can hold only so many ideas at once, so there's a limit to the span of time we can perceive. "Were our senses altered, and made much quicker and acuter, the appearance and outward scheme of things would have quite another face to us."

William James took up the idea. Suppose your senses are altered by hashish, he wrote in 1886; you might have a temporal experience akin "to the condition of Von Baer's and Spencer's short-lived beings. . . . The condition would, in short, be exactly analogous to the enlargement of space by a microscope; fewer real things at once in the immediate field of view, but each of them taking up more than its normal room, and making the excluded ones seem unnaturally far away." In 1901 H. G. Wells wrote a short story called "The New Accelerator," about the invention of an elixir that speeds up the body and its perceptions by a thousand times. Let a drinking glass drop and it would appear to hang in midair; people on the street would be "smitten rigid, as it were, into the semblance of realistic wax," he wrote. "We shall manufacture and sell the Accelerator, and, as for the consequences—we shall see."

Although we rarely appreciate the fact, humans function in and rely on several different timescales simultaneously. The average human heart beats once per second. Lightning strikes in a hundredth of a second. A home computer executes a single software instruction in nanoseconds, or billionths of a second. Circuits have switching times in picoseconds, or trillionths of a second. Several years ago, physicists managed to create a pulse of laser light lasting only five femtoseconds, or five quadrillionths (5×10^{-15}) of a second. In everyday photography, a camera's flash can "stop time" at about one one-thousandth of a second—fast enough to freeze the swing of a baseball batter, if not a speeding fastball. Likewise, the femtosecond "flashbulb" enabled scientists to observe phenomena never before seen in freeze-frame: vibrating molecules, the binding of atoms during chemical reactions, and other ultrasmall, ultra-fleeting events.

The femtosecond pulse has evolved into a powerful tool. It is superb for drilling tiny holes; its energy is deposited so quickly that there's no time for the surrounding material to heat up, so there's less mess and inefficiency. Also, given the speed of light—just under three hundred million meters per second—a pulse of light one femtosecond long has a physical length of only about a thousandth of a millimeter. (In contrast, a pulse of light one second long would stretch four-fifths of the way from Earth to the moon.) Think of them as tiny smart bombs: they can be focused to strike just below the surface of a transparent material without actually piercing it. Femtosecond pulses are being used to etch optical waveguides inside panes of glass, a development that could revolutionize data storage and telecommunications. Femtosecond researchers have developed a new method of laser eye surgery that operates directly on the cornea without damaging the tissue above it. "It's a way of putting your hand inside biological materials, and doing so with very little energy," Paul Corkum, a physicist at the University of Ottawa, told me.

But ultrafast is not good enough. All kinds of important things can happen between one quadrillionth of a second and the next, and if your flashbulb is too slow, you'll miss out. So scientists have been pressing on, punching the clock, hurrying to create even tinier windows of time through which to study the physical world. A few years ago an international team of physicists, including Corkum, finally succeeded in breaking the so-called femtosecond barrier. With a complex, high-energy laser, they generated a pulse of light little more than half a femtosecond long—650 attoseconds, to be precise. The attosecond (10^{-18} second) has long existed as a theoretical entity but this was the first time anyone had actually encountered it. It's a newfound slice of time—a tiny one but with gargantuan potential. "This is the real timescale of matter," Corkum said. "We're gaining the ability to look at the microworld of atoms and molecules on its own terms."

No sooner had the physicists caught an attosecond pulse than they demonstrated its usefulness. They aimed it and a longer pulse of red light into a gas of krypton atoms. The attosecond pulse excited the krypton atoms, kicking electrons free; then the red-light pulse hit the electrons and took a reading of their energy. By adjusting the delay between the two pulses, the scientists were able to take a very precise measurement—within a matter of attoseconds—of how long it takes the electron to decay. Never before had electron dynamics been studied on so short a timescale. The experiment set the physics world buzzing. "Attoseconds will give us a new way to think about electrons," Louis Di-Mauro, a physicist at Brookhaven National Laboratory, told me. "They become a new probe of matter that will then be applied across the sciences. The age of attophysics has begun."

Of course, one day, perhaps not so very far in the future, even the attosecond will fail to satisfy. To probe the activity of the atomic nucleus, where the natural timescale is several orders of magnitude faster, physicists will have to break into the realm of zeptoseconds, or sextillionths of a second. In the meantime they will have to manage with the little free time they've gained already. One can imagine them getting carried away, filling up their hard drives with electron home videos, clogging the airwaves with attosecond videos that seem to yawn for seconds—for eternity, basically. Corkum is confident this won't happen: "In practice, we're only looking at a reasonable period of time." In small time, just like in the big time, viewer boredom still sets the limits. "My brother-in-law recently sent some movies of their baby," Corkum said. "It was fun at first, but after fifteen minutes—wow, that's a lot of time."

When I was younger and had more time, I liked to lie on the grass in the summer, close my eyes, and try to count how many sounds I could hear simultaneously. Over there, the buzz of cicadas. Above, the high-altitude roar of a jet. Behind me, the leaves of a tree rustling in the breeze. Some sounds were a steady presence; others, like a blue jay's call, came and went. I found that I could hold four or five in mind before one fell out, at which point I would cast around for another one, like a juggler who has dropped a ball and with one hand gropes around for another ball to replace it. Before long I'd settle into the constant plus and minus of the process and the challenge became less about the number of sounds I could keep aloft and more about the internal space they occupied and the least effort needed to keep them in motion.

This was relaxing but also was my way of measuring . . . well, I wasn't sure exactly. The span of my attention? The boundary of my awareness? In retrospect, it's clear that I was trying in my rudimentary way to quantify the present moment, an endeavor with a long history. Even before William James (by way of E. R. Clay) landed on the notion of the specious present, scientists had mostly accepted the notion that the psychological present has some actual duration, and much effort was spent trying to quantify it. Exactly how long, or short, is now?

One way to measure the present was to count how many mental items could fit in it. Rhythms were a useful gauge. Imagine a series of beats that produce a rhythm like this: *tiketta-tik-tik-tik tiketta-tik-tik-tik,* and so on. If the individual beats arrive too slowly or too quickly, the rhythm isn't discernible; only at some intermediate range of speeds— so many beats per second or minute—do the beats fuse in your mind to form a whole. Or, to put it differently, the rhythm emerges only when a sufficient number of individual beats (but not too many) land within a brief, slightly variable period of awareness. The German physiologist

Wilhelm Wundt called this "the scope of consciousness," or *Blickfield*—the short interval in which disparate impressions fuse into a sense of now. In the eighteen-seventies he embarked on an effort to measure its parameters. In one experiment, he played a series of sixteen beats—eight pairs of two beats—at a rate of one to one-and-a-half beats per second, defining a *Blickfield* between 10.6 and 16 seconds long. He played the series twice for subjects: once, and then, after a brief pause, again. His subjects immediately discerned a rhythm and also recognized that the two rhythms were the same. If he added a beat to the second series, or dropped a beat, the listener immediately noticed it, even without counting individual beats. They were aware of an overall pattern; each rhythm "is in consciousness as a whole," Wundt remarked. He sped things up, so that twelve individual beats came every one-half to one-third of a second, and still the subject could discern a rhythm or "whole" and compare it to another. By this measure the discernible *now* lasts somewhere between four and six seconds long. As many as forty beats could be recognized at once, provided they arrived in five groups of eight apiece, at a rate of four beats per second. (That would put the range of consciousness at ten seconds.) The briefest perceivable duration consisted twelve beats—three groups of four beats, at three beats per second—and lasted four seconds.

By other measures, now could be far shorter. In 1873 the Austrian physiologist Sigmund Exner reported that he could hear two successive snaps of an electric spark if one followed the other by as little as one five-hundredth of a second. Whereas Wundt's subjects were assessing the contents of a filled moment, Exner was marking the borders of an empty one. And the size of this now, Exner found, very much depends on which senses engage it. Hearing gave access to the smallest perceptible interval (0.002 seconds). Eyesight was slower: if Exner watched two successive sparks that sat slightly apart from each other, he could correctly discern which one came first only if they occurred more than 0.045 seconds (slightly less than one-twentieth of a second) apart. If the task involved hearing a sound and then seeing a light, an even longer interval (0.06 seconds) was needed to discern their order. The smallest perceptible interval for the reverse task, from sight to hearing, was longer still, 0.16 seconds.

A few years earlier, in 1868, the German physician Karl von Vierordt had offered another measure of now. In his experiments, subjects listened to an empty interval—often marked by two ticks on a metronome—and would try to reproduce it, typically by pressing a key that made a mark on a rotating drum of paper. Sometimes the interval to be reproduced was marked by eight beats of a metronome rather than two, or the two beats might be delivered on the subject's hand with a small steel point. In reviewing the data Vierordt noticed something curious: durations lasting less than a second or so were typically judged to be slightly longer than they actually were, whereas longer durations tended to be underestimated. Somewhere in between was a brief duration that the subjects judged accurately. Through many experiments Vierordt strived to pinpoint this short interval at which one's sensation of time corresponds precisely to physical time. He called this the indifference point. It varied from one observer to another, but on average, later researchers claimed, it was remarkably constant at about 0.75 seconds long.

It's now clear that the finding disguised several methodological flaws. For one, nearly all of Vierordt's experimental data were derived from just two volunteers, Vierordt and his doctoral student. Nonetheless, the indifference point was widely embraced as significant. Wundt and others ran their own indifference point experiments to further quantify and articulate it; their values often circled around three-quarters of a second, though some ranged as low as a third of a second. Evidence for the indifference point largely dissolved under later scrutiny, but for a moment, at least, scientists seemed to have identified a psychological unit of time—"some absolute duration," one historian has written, that "is always available to the mind as a standard." This duration, whatever its exact size, stood as a proxy for consciousness; it was the smallest possible moment of direct human awareness.

The exact size of now would be prodded and parsed well into the twentieth century. Scientists these days tend to draw a distinction between two concepts. One is the perceptual moment, a fleeting yet quantifiable duration that is defined as the largest interval between two successive events, such as a pair of sparks, which are nonetheless perceived as simultaneous. The other is the psychological present, a slightly

longer period in which a single event, such as a drum roll, seems to unfold. The former might be ninety seconds, or 4.5 milliseconds, or somewhere between a fifth and a twentieth of a second long, depending on
whom you ask and how you measure; the latter might be two to three
seconds, or four to seven seconds, or no more than five seconds long.
At least one group of cognitive scientists has proposed the existence
of a time quantum, "an absolute lower bound of temporal resolution,"
which they place at roughly 4.5 milliseconds.

By the time James published *Principles of Psychology*, in 1890,
the size of now was basically settled, in his view. "We are constantly
conscious of a certain duration—the specious present—varying in
length from a few seconds to probably not more than a minute," he
wrote. Additional investigation—further "starving out, and harass-
ing"—was undignified: "There is little of the grand style about these
new prism, pendulum, and chronograph-philosophers. They mean
business, not chivalry." James looked on the new phase of German
research as "a microscopic psychology" that "taxes patience to the ut-
most, and could hardly have arisen in a country whose natives could
be bored." There were better things to do with one's time than peck it
to death.

Whatever such experiments revealed about our "time sense," they
were testimony to the increasing accuracy of mechanical timekeepers.
Scientists had long been intrigued by the "animal spirits" or "nerve actions" that animated muscles and enabled movement, cognition, and
the perception of time. But neural impulses, as they're called today, can
travel as fast as four hundred feet per second, or two hundred and fifty
miles per hour—far too fast for eighteenth-century technology to detect. As far as science was concerned, an action followed instantaneously
from the thought that provoked it. But in the nineteenth century, advances in temporal measurement—pendulum clocks, chronoscopes,
chronographs, kymographs, and other devices borrowed largely from
astronomy—provided access to new timescales: tenths, hundredths,
even thousandths of a second. Instruments designed to probe the cos-

mos were applied to the study of physiology, and they opened a window of time large enough to reveal the unconscious.

Until relatively recently, with the embrace of atomic time and a universal time standard so advanced that it must be disseminated by newsletter, the time on our clocks and watches came to us by way of astronomical observatories, which gleaned it from the stars. Imagine a line that passes overhead and connects due north and due south; wherever you are, the sun crosses that line—the celestial meridian—every day precisely at your noon. (Solar noon is defined as the moment when that occurs.) At night stars cross, or transit, the meridian at equally precise times; astronomers came to track these star transits closely. One could set a clock by them—and watchmakers and clock owners did, at first by pestering the local astronomer directly and later by subscribing to some form of observatory-approved "time service." In 1858 an observatory was built in Neuchâtel, Switzerland, expressly to provide accurate time to the timepiece industry. "Time will be distributed to homes, like water or gas," Adolph Hirsch, the observatory's founder and chief astronomer, boasted. Local manufacturers could drop off their clocks and watches at the observatory to be tested, calibrated, and officially certified. Clockmakers farther away received a daily time signal by telegraph. By 1860 every Swiss telegraph office received its time from Neuchâtel, thereby establishing what Henning Schmidgen, a historian and professor of media theory at Bauhaus University, in Weimar, has called "a vast landscape of standard time."

Of course it isn't precisely noon, or any particular time, everywhere on earth simultaneously. Our world turns, so the sun doesn't shine on all of us at once; midday in New York is midnight in Hong Kong. As you travel east, sunrise and sunset (and noon) occur slightly earlier relative to your starting point; they occur later as you travel west. For every fifteen degrees of longitude east or west (out of three hundred and sixty degrees total), noon occurs an hour earlier or later, respectively. With a telescope and a clock you can map out the world. Suppose you're an astronomer at the Greenwich Observatory, which lies on longitude 0°; if you know the time at which a certain star crosses your meridian, you can accurately predict the moment that it will cross the meridian at, say, longitude 35° W, halfway across the Atlantic. Now, instead, place yourself on that ship and with your telescope and a clock determine the exact moment that the same star crosses your meridian. If you also know the exact time at which the same star crosses the meridian in Greenwich, you can calculate your longitude from the difference in the transit times. This method of reckoning was central to British exploration in the sixteenth and seventeenth centuries. It drove the invention of accurate seaworthy clocks and prompted the construction, in 1675, of the Royal Observatory in Greenwich, the first observatory built to provide a solid baseline against which far-flung ships could fix their whereabouts.

The process of establishing local time from a stellar transit was painstaking. As the appointed moment approached, the astronomer would glance at the clock, note the time to the nearest second, then peer into his telescope. Spanning his field of view was a series of evenly spaced vertical lines, which often were marked on the telescope with spiders' thread. Before long the star—a bright point of silver light, in a halo of color—would swim into view. As the astronomer counted seconds aloud, or listened to the clock or sometimes to a metronome beating out the seconds, he noted exactly when the star crossed each line, especially the middle one, which represented the meridian. This entailed visually fixing the star's location on the beat just before it crossed the line and on the beat just after; remembering both positions; comparing them; and expressing the difference—the exact moment of crossing—all in tenths of seconds. Transit times could be compared across days or weeks. Since the star was always punctual, any deviation from its expected schedule was safely blamed on the clock, which would be reset accordingly.

The technique was assumed to be accurate within two-tenths of a second, but it was fraught with error. Telescopes differed in clarity from one observatory to the next, and not every observatory clock beat as steadily as the next one or was as insulated from outside noise and vibration. A star might be unusually bright or dim; it might tremble in hidden air currents; it might vanish under a cloud at the critical moment. More insidious was a type of human error that became known in astronomy as the "personal equation." In 1795 the royal astronomer at Greenwich noted that he had fired his assistant because the man's stellar transit times were consistently a second slower than the ones he recorded himself: "He fell into some irregular and confused method of his own." Yet it soon became clear that no two observers recorded precisely the same transit time; everyone had a personal equation. For the next fifty years astronomers across Europe measured and compared their errors in a fruitless effort to factor them out.

The fault lay in human physiology: it was "an unfortunate characteristic of the astronomers' nervous system," Hirsch concluded in 1862. A decade earlier, experiments by Hermann von Helmholtz, the German physicist and physiologist, had revealed that perception, thought, and action aren't instantaneous after all; the speed of human thought is finite. By applying a weak electric shock to different parts of a volunteer's body, he could measure how long it took the subject to respond to the stimulus, which the subject did by moving his head. Response times varied wildly, but Helmholtz broadly calculated that human nerve signals travel at about a hundred and twenty feet per second, far slower than the nine-million-miles-per-hour estimate given by some investigators. Helmholtz compared human nerves to telegraph cables that "send reports from the furthest borders of a state to the ruling center." Such a transmission takes time—time to become aware of a stimulus, time to carry out a response, and, in between, the time required "by the brain for the processes of perceiving and willing," Helmholtz wrote. He estimated that the perceiving-and-willing step took one-tenth of a second.

Hirsch, the astronomer, referred to this interval as "physiological time," and he suspected that it was responsible for the personal equation. He pursued a series of experiments to clarify the matter. In one, a steel ball dropped loudly onto a board; at the sound, the test subject tapped a telegraph key. Hirsch measured the time between the sound and the keystroke with a chronoscope, a device capable of measuring time intervals to within a thousandth of a second, and landed on a nerve speed roughly half of what Helmholtz had calculated. The chronoscope had been invented some years earlier by Matthias Hipp, a clockmaker who later became one of Hirsch's research volunteers, to measure the velocity of shotgun pellets and falling objects. Hipp subsequently became the director of the Swiss telegraph service; in 1860 he retired to start his own telegraph company in Neuchâtel, in part to supply equipment to Hirsch's new time-transmission business. Hirsch ran experiments with a contraption that showed artificial stars crossing the lines of a meridian telescope. He found that the personal equation didn't just vary from person to person. It varied from observation to observation, over the course of the day, and over the year; it varied depending on the brightness of the star and the direction in which it was moving. If you recorded the transit time by anticipating when it would cross the meridian line, rather than waiting until it had actually done so, that would alter the personal equation too.

Astronomers soon learned to factor out the personal equation by depersonalizing the process of astronomical observation. The eye-and-ear method was replaced by the electrochronograph, a rotating drum of paper attached directly to the clock. The astronomer noted a star transit, pressed a key, and marked the paper, eliminating the need to view or think about the clock and, with it, the uniquely personal time delay. Now two astronomers could compare their errors objectively on the same clock. Seated miles apart in separate observatories, they could simultaneously record the same stellar transit, using time from a single clock shared via telegraph (after factoring out the telegraph transmission time), and calculate out their differences.

But the personal equation had left a mark; the study of time had spread from astronomy into physiology and psychology. Hirsch's 1862 paper, with its reference to "physiological time," was translated from German and circulated more widely among scientists. His experimental design for studying astronomers served as the model for some of Wilhelm Wundt's subsequent experiments on the temporal span of consciousness. Interest grew in the study of response times. In 1926 and 1927, Bernice Graves, a football coach pursuing his master's in psychology at Stanford, conducted a study of the Stanford football players' reaction times, under the guidance of psychologist Walter Miles and Glenn "Pop" Warner, the team's coach. Central to the endeavor was a timing contraption that Miles had invented but which would have looked familiar to Hirsch. Miles called it a "multiple chronograph," and it could be connected to seven linemen simultaneously to measure how quickly they charged off the line after the quarterback's signal to snap the ball. At the time, there was considerable debate over which method of signaling was best. An audible signal—in which the quarterback calls out a series of numbers detailing the play to his teammates, followed by a loud "Hike!"—was clearly superior to a visual one, as the offensive linemen could keep their eyes on the defenders lined up against them. But should the linemen be surprised by the "hike" or, as part of the audible code, be forewarned that it was coming? Should the cadence of the signal be even or varied? Graves tested out all the variables with Miles's timing device. In a three-point stance, each lineman rested his head against a trigger release; at the signal, the lineman moved, which triggered a golf ball to drop and leave a mark on a rotating drum of paper. Reaction time could be measured in thousandths of a second. Graves found that the players moved off the line more uniformly when the signal was unexpected and nonrhythmic. But when the signal was anticipated and rhythmic the players charged forward more quickly, by as much as a tenth of a second—the time otherwise required, more or less, for a person to think. "Snappy, precise action in unity is the strength objective toward which the coach works and the men train," Miles noted. "The effort is to make the eleven individual nervous systems into one well-integrated, powerful machine."

Walking back from the deli to my office one day after lunch, I glance up at a clock that sits on a high pedestal outside the bank. It somewhat resembles a giant ship's compass, and I'm suddenly made aware of the clock's quiet efforts to orient me, in more ways than one.

In fact this clock—or the clock on my cell phone or on my bedside table, or the wristwatch I sometimes wear—has several things to tell me about time. At its most basic, a clock is a timer; it situates me with respect to a little while ago and some time coming up. "If I take out my watch," the philosopher Martin Heidegger remarked, "then the first thing I say is: 'Now it is nine o'clock; thirty minutes since that occurred.

In three hours it will be twelve.'" In other words, a clock orients with respect to the past and future; its aim, as Heidegger put it, "is to determine the specific fixing of the now"—*now* being an ever-moving target.

But that information alone is only marginally helpful; my *now* is a ship adrift unless it refers to established landmarks. One such landmark is the sun: a clock tells me where I am in the day. A nightstand clock that reads 2:00 p.m., when I can clearly see it's the middle of the night, is doing something seriously wrong; it is out of step with the rotation of the earth. Also, implicitly, the clock tells me where (or rather, when) I am in relation to clocks other than the one I'm currently viewing. If the bank clock says 2:00 p.m. when I walk past it on my way to catch a 2:15 train, I would like not to reach the station in five minutes and find that the clock there reads, say, 2:30 p.m. and that I've missed my train. We expect our clocks to be in sync with one another as well as with the planetary day at large. My now should be your now, even if you're on the other side of the world.

This expectation is ingrained in modern digital life but it wasn't always. In the nineteenth century, Europe, the United States, and the rest of the world were struggling to emerge from what the historian Peter Galison has called "the chaos of uncoordinated time." Astronomy enabled every town that wanted an accurate clock to have one. But a local clock sufficed only as long as no one went anywhere. As railroads spread and offered faster movement across wider distances, travelers discovered that the time in one city rarely agreed exactly with the time in another. In 1866, when it was noon in Washington, D.C., the official local time in Savannah was 11:43; in Buffalo, 11:52; in Rochester, 11:58; in Philadelphia, 12:07; in New York, 12:12; and in Boston, 12:24. There were more than two dozen distinct local times in the state of Illinois alone. In 1882, when William James sailed to Europe to meet with leading psychologists and to try to make progress on his book, he left behind a nation with anywhere from sixty to a hundred local standard times.

For the sake of convenience, to simplify railroad timetables, and to keep trains from crashing, an effort arose to coordinate clocks across and among cities, using the telegraph to exchange time signals. Simultaneity became a distributed commodity; the landscape of time changed from minutely granular terrain into wide, more regular tracts of now. James returned to the United States in the spring of 1883; later that year, at precisely noon on Sunday, November 18th, the government officially reduced the nation's time zones from dozens to just four. The event became known as "the Day of Two Noons," as it required half the population in each new zone to set their clocks back slightly and experience noon once more. "Those in the eastern half of the zone are, as it were, 'living a little of their lives over again' but those on the other side are thrown, some of them as much as half an hour, into the future," the *New York Herald* remarked.

At the turn of the century, with tremendous political effort, the time-keeping systems of the world were persuaded to similarly coordinate with one another. Invisible lines were drawn and twenty-four evenly spaced time zones were established on the globe. *Now* was specifically fixed for everyone on earth. The French mathematician Henri Poincaré, a leading voice in the movement, noted that time is nothing more than "a convention." In French, Galison writes, *convention* carries two meanings: a consensus, or a meeting of opinions, and a convenience. *Now* is whenever we all agree it is, in order to make our shared lives easier.

This was a novel concept. Since the seventeenth century, physicists had mostly followed Isaac Newton's belief that time and space are "infinite, homogenous, continuous entities, entirely independent of any sensible object or motion by which we try to measure them." Newton added, "Absolute, true, and mathematical time, of itself, and by its own nature, flows uniformly on, without regard to anything external." Time was inherent in the fabric of the universe, a stage unto itself. With the twentieth century, time became strictly quotidian; it exists only in measurement. Einstein was blunt: time is no less and no more than "what we measure with a clock."

So when I wake at night and decline to look at my bedside clock, I'm engaging in a form of protest. The world of time is by definition social, a common accord by which peoples and nations navigate their troubles and needs. My clock offers to reveal *now* to me—to fix it specifically with numbers—but only if I sign on to the universal convention. In the middle of the night, or whatever time it is, I want my time to be mine alone.

I'm aware that this is a delusion. Every living body—my own, that of the nebulous deep-sea jelly or the microbial plaque that grows on my teeth while I sleep—is an organization of parts: cells, cilia, cytoskeletons, organs, and organelles, down to the heritable bits of genetic data that enable some aspect of every individual to persist across the ages. To be organized is to communicate—to arrange which part does what when and in what order. Time is the conversation through which our parts create a whole greater than their sum. I can ignore this chatter in the middle of the night and drift alone, for a little while, but only as long as I don't press too deeply for a definition of "I."

The industrialization of the late nineteenth century is often characterized as a period of dehumanization: labor grew increasingly rote and mechanical, workers became like cogs in machinery. But as the twentieth century approached, the city as a whole underwent an opposite transformation, taking on the characteristics of a living body. Its borders widened, its mass swelled with residents; networks of pipes and wires grew to supply demand. "The big city looks more and more like a perfect organism, having its nervous system . . . , its blood vessels, its arteries and veins distributing gas and vital water from one end to the other," a Berlin schoolbook noted, in 1873. "Only if the streets are torn up for repairs does one perceive these hidden spirits, which radiate their mysterious effectiveness far below."

Meanwhile, the study of living organisms grew more technical. To understand the workings of what the German physiologist Emil du Bois-Reymond called the "animal machine"—respiration, muscle movements, nerve signals, the flow of blood and lymph, the beating of the heart—one needed mechanisms: belt pulleys, rotating engines, gas power. One laboratory, aided by two motors running in the basement, studied "the disturbances that take place in animals"—chiefly frogs and dogs—"as a result of spinning." Cats and rabbits were dissected alive to deduce how the organs function; a bellows was required to keep the animal breathing. But pumping the bellows was hard work for a human assistant, so by the eighteen-seventies, the job had been given to mechanical pumps, which enabled an animal to breathe uniformly and exactly, like clockwork. Here in the physiology factory, the historian Sven Dierig has noted, "the first living organism that was part machine and part animal was created and brought into use for scientific purposes." Fine timing made it possible.

This was the golden age of automata: mechanical men, animated by intricate interior clockworks, that could pull a carriage, recite the alphabet, draw a picture, or write a name. To Karl Marx, factories themselves were automata: "Here we have, in the place of the isolated machine, a mechanical monster whose body fills whole factories, and whose demon power, at first veiled under the slow and measured motions of his giant limbs, at length breaks out into the fast and furious whirl of his countless working organs." The conflation of metaphors only deepened the mystery: what exactly distinguishes humans from clockwork, a mind from a moving body? How does the living mechanism give rise to consciousness? And where does that evanescent something—the homunculus in us, the soul, the spirit beneath the street—lie hidden? "Even though it is highly improbable, and will always remain a vain wish to put the Homunculus together, scientists have already taken some important steps in that direction," Wilhelm Wundt noted in 1862. The previous year, a French anatomist named Paul Broca had discovered that a filament of cortex in the left frontal lobe of the brain is essential to human speech and memory. Thomas Edison was fascinated by this finding. "Eighty-two remarkable operations upon the brain have definitely proven that the meat of our personality lies in that part of the brain known as the fold of Broca," he said in 1922. "Everything we call memory goes on in a little strip not much more than a quarter of an inch long. This is where the little people live who keep our records for us."

The manufacture and study of time was also growing more industrial. In 1811 Greenwich Observatory had a single employee, the Astronomer Royal. By 1900 there were fifty-three people on staff, half of them involved solely in carrying out calculations; these employees were called "computers." In the new psychology laboratories, telegraphs, chronographs, chronoscopes, and other highly accurate timekeeping devices were employed to measure response times and temporal perceptions. And astronomers and psychologists alike complained about the din—the rumble of machinery, the clatter of traffic, the noise and vibrations from outside that seeped in, shook, rattled, distracted, and disrupted.

Often the loudest noises came from the laboratories themselves. By now psychologists recognized that their subjects' estimates of a duration—how long a chime seems to last, for instance—varied with attention. Focus was critical, but the clicks and whistles from the time-keeping machinery used for the studies could be as distracting to subjects as the noise from outside. "I hear the sound of the chronoscope," one research volunteer complained. "I can't get rid of it." The scientists struggled to get away from themselves. They built quieter tools and quieter surroundings; subjects were placed in a room separate from the experimental equipment and linked to the researcher by telegraph and telephone lines. The time laboratory, strung with cables and wires, more and more resembled the city that it sought to escape—and the cerebral networks that it sought to decipher. These days we talk casually of the brain sending "signals" that the nerves "transmit." That metaphor entered physiology in the nineteenth century and was borrowed directly from the telegraph industry.

Eventually, perhaps inevitably, the isolation booth was devised. The Yale psychologist Edward Wheeler Scripture offered a builder's guide: a room within a room in the middle of the building, with air-tight brick walls resting on rubber supports and sawdust filling the space between the walls. You enter through heavy doors. "The room should be furnished and lighted exactly like a comfortable room in the evening. All wires and apparatus should be concealed. The person entering it should suppose it to be only a reception room. He is to believe that he is merely on a visit."

Imagine yourself in a windowless phone booth, the lights turned off, an observer left in dark and utter silence. Or nearly so; there is one last noise that Scripture can't eliminate. "Alas! We have let in a sad source of disturbance, namely, the person himself," he laments. He describes his own experience: "My clothes creak, scrape, and rustle with every breath; the muscles of the cheeks and eyelids rumble; if I happen to move my teeth, the noise seems terrific. I hear a loud and terrible roaring in the head; of course, I know it is merely the noise of the blood rushing through the arteries of the ears . . . but I can readily imagine that I possess an antiquated clockwork and that, when I think, I can hear the wheels go 'round."

What Just Happened

I t's 5:28 a.m. on Wednesday, April 18th, 1906, and William James is
wide awake, as usual. James is living in Palo Alto, teaching at Stanford for a semester. "It was a simple life," he wrote in May to his
friend John Jay Chapman. "I am glad for once to have been part of the
working machine of California."

Suddenly his bed begins shaking violently. James sits up and immediately is thrown flat and shaken "exactly as a terrier shakes a rat," he
recalls later in another letter. It's an earthquake. James has been curious
about earthquakes and now here one is; he is almost giddy. But there's
hardly time for that. The bureau and chiffonier tip over, the plaster walls
crack, "an awful roaring" fills the air, he notes. And then, in an instant,
all is still again.

James is unscathed. The earthquake was "a memorable bit of experience, and altogether we have found it mind-enlarging," he tells
Chapman. He recalls the experience of a Stanford student who had
been sleeping on the fourth floor of a dormitory when the quake shook
him awake. As he rose, books and furniture fell to the floor, and he was
thrown off his feet. Then the chimney collapsed through the center
of building, and the books and furniture and the student were carried
with it down a ragged rabbit hole. James writes, "With an awful, sinis-
ter, grinding roar, everything gave way, and with chimneys, floor-beams,
walls and all, he descended through the three lower stories of the building into the basement. 'This is my end, this is my death,' he felt; but all
the while no trace of fear."

· · ·

I am falling, that much I know. The sky, when last I noticed, was a peerless blue expanse. Now it grows ever so slightly larger and more distant as I fall away from it, backward, toward earth.

I also know, because I did the math in advance, that a fall from a hundred-foot height—in my case, from the Zero Gravity Thrill Amusement Park's Nothin' but Net 100-foot Free Fall attraction, which is little more than a scaffold tower and a pair of nets above a dusty lot in Dallas—will last less than three seconds. I don't know where I am in that span, only that my fall has begun and is not yet complete.

It's often said that time slows down during moments of trauma and extreme stress. A friend crashes his bike and years later vividly recalls the dilated moments: his hand extended to break the fall, a truck skidding to a halt just inches from his head. A man's car stalls in the path of an oncoming train and, with a clarity of thought and action that astounds him, realizes that he has just enough time before the collision to pull his daughter into the front seat and shield her with his body. Research volunteers who watch a stressful video of a bank robbery report the event taking longer than it actually did. Novice skydivers overestimate the length of their first jumps roughly in correlation to their level of fear.

I am here now, falling through the present, to see if time will slow for me too. Can I accomplish more during this dilated present—react more quickly, perceive my surroundings in more detail? How does one even begin to study such a thing? Scientists who attempt to grapple with such questions inevitably encounter a conundrum: *When* should this allegedly expanded now be studied? Right now in the perhaps inaccessibly brief moment of its occurrence? Or afterward, when it becomes difficult to distinguish what actually occurred from the unreliable memory of it? One can't venture far in thinking about time without tumbling into one of the deepest questions in the literature: how long is the present, and where is the human mind with respect to it? As the historian of psychology Edward G. Boring (whose writing in fact is quite compelling) once put it, "At what time in a time does one perceive a time?" Augustine, of course, had an answer: "We can only hope to measure it as it passes by, because once it has passed . . . it will not exist to be measured."

Which brings me to now, or whenever we are. Some psychological studies have concluded that what we call "the present" is framed by blinks of one's eyes, which would have it lasting about three seconds. I doubt that my own eyes are a reliable metric. They may be blinking rapidly; they may not be blinking at all—who notices that sort of thing most of the time? The wind must be howling past my ears as I fall too, but if so I can't hear it. Three seconds would seem like hardly enough time to think about anything, and what I recall later might well differ from what I sense right now. For the moment, all I sense for certain is that I am falling at ever greater speed.

When David Eagleman was eight years old, he fell off a roof. "My memory of it is so clear," he told me. "There was this tarpaper hanging over the edge—I didn't know the word 'tarpaper' at the time—that I thought was the edge, so I stepped on it. Then I was going."

Eagleman distinctly remembers the sensation of time slowing down during his fall. "I had a series of thoughts that were motionless and clear, like, 'I wonder if I have time to grab on to that tarpaper,'" he said. "But I sort of knew that it would probably tear. And then I realized I probably wouldn't have time to reach it anyway. So then I was heading toward the brick floor and watching it come toward me."

Eagleman was lucky; he was unconscious for a few moments and walked away with only a broken nose. But he remained fascinated by the experience of slowed time. "All through my teens and twenties, I read a lot of popular physics about time and contraction: *The Universe and Dr. Einstein*, that sort of stuff. I found that interesting, that time is not a constant thing."

Eagleman is a neuroscientist at Stanford, where he studies, among other things, the perception of time; it's a recent appointment, and for many years previously he worked at the Baylor College of Medicine, in Houston. Researchers who study time come with different specialties. Some focus on circadian clocks, the twenty-four-hour biological rhythms that govern our days. Others study "interval timing," the brain's ability to plan, estimate, and make decisions over time intervals lasting from roughly one second to several minutes. A much smaller group of scientists, including Eagleman, study the neural basis of time across milliseconds, or thousandths of a second. A seemingly slight window of time, milliseconds in fact govern many of our basic human activities, including our ability to produce and understand speech, and underpin our intuitive sense of causality. To grasp the instants—and how the human brain perceives and processes them—is to grasp the fundamental units of human experience. But whereas the workings of the circadian clock have been closely described in the past two decades, researchers have only begun to argue over how the brain's "interval timer" works, its whereabouts in the brain, and whether a single-clock model even applies. The millisecond-timing clock, if there is such a thing, is an even greater enigma, in part because the tools of neuroscience have only recently attained sufficient precision to probe timing activity at that fine scale.

Eagleman brims with energy and with ideas that cross the standard academic boundaries. When I first met him, he had just published a novel, *Sum*, and had begun conducting a series of seemingly slight yet deeply provocative experiments, including the one staged at the Zero Gravity Thrill Amusement Park's Nothin' but Net 100-foot Free Fall attraction, to explore how time slows down and why. Since then, he has written five books and hosted a public television series about the brain, been the subject of magazine profiles, including one in *The New Yorker*, and given a popular TED talk. He moved to the Bay Area in part to develop two entrepreneurial ideas: a vest for the deaf that translates sonic vibrations into tactile sensations, enabling the wearer to hear sounds somewhat in the way that braille enables a blind person to read; and a smartphone and tablet app that, through a series of cognitive games, helps reveal if the user has suffered a concussion.

It's the sort of activity and attention that can invite skepticism and professional envy, particularly from neurobiologists, whose research deals more directly with the brain's wetware and doesn't always provide the same sweeping clarity, or excitement, that cognitive scientists can generate. "I am impressed and amused by David's work," one leading time researcher told me. But Eagleman's colleagues also note that his studies have made a genuine mark on the field. Once when I visited him at Baylor, he had invited Warren Meck, a neurobiologist at Duke University and the authority on interval timing, to give a talk to his department. Meck, a quietly intimidating figure with a wry smile, introduced his subject by saying, "I'm the wave of the past; I'm father time. David is the wave of the future."

Eagleman grew up in Albuquerque, New Mexico, as David Egelman, the second son of a psychiatrist (his father) and a biology professor (his mother). (*Egelman* is pronounced "Eagleman," but so many people saw it and mispronounced it—or misspelled it after hearing him pronounce it correctly—that he later changed the official spelling.) At home, conversations about the brain were "part of the background radiation," he said. He started college at Rice University with a double major in literature and space physics. He did well but dropped out after sophomore year, bored and dispirited. He studied at Oxford for a semester, then lived in Los Angeles for a year, where he worked for production companies reading scripts and planning lavish parties that he wasn't yet legally old enough to attend. He returned to Rice, to finish his literature degree, but soon began spending his free time at the library reading everything he could find about the brain.

When senior year came he applied to film school at U.C.L.A. A friend suggested that he consider becoming a neuroscientist, despite the fact that he'd taken no biology since tenth grade, so he applied to Baylor's graduate program in neuroscience. He emphasized his undergraduate coursework in math and physics and submitted a long paper he'd written, based on his extracurricular reading, that summarized his personal theories about the brain. ("In retrospect, it's totally embarrassing," he said.) As a backup plan he thought he might become an airline attendant, on the premise that he would "fly to different countries and write novels."

He was a semifinalist at U.C.L.A. but went to Baylor. During his first week at grad school he had anxiety dreams, including one in which his adviser told him that the acceptance letter was a mistake and was intended instead for someone named David Engleman. But he did well in graduate school and went to the Salk Institute, in San Diego, for his postdoctoral research. Before long he was recruited to Baylor with funding to run a small laboratory, the Laboratory for Perception and Action. It was one of several labs lining a long hallway in a labyrinth of corridors, with room for several cubicles for his grad students, a meeting table, a small kitchen, and his own interior office. When I asked, he described his own relationship to time as "mixed." He is often late with deadlines, he said, writes while standing up, and dislikes napping. "If I accidentally fall asleep for thirty-five minutes," he said, "I wake up and think, 'I'll never get those thirty-five minutes back.'"

As a postdoc, Eagleman worked on a sprawling computer simulation of how neurons in the human brain interact. Time perception was not on his radar until he encountered the flash-lag effect, one of the lesser-known sensory illusions that have engaged psychologists and cognitive scientists over the years. In his office he showed me a copy of *The Great Book of Optical Illusions*, by Al Seckel, which described hundreds of illusions, including one of the oldest, the "motion aftereffect," sometimes called the waterfall effect. Watch a waterfall for a minute or so and then look away; everything in sight will now seem to be crawling upward. "In physics, motion is defined as a change in position over time," Eagleman said. "But it's not that way in the brain: you can have motion without change in position."

Eagleman likes illusions. With each one it is as if, while enjoying the theater of the senses, you've caught sight of a stagehand moving the sets around; it's a gentle reminder that our conscious experience is a manufactured affair and that, with marvelous reliability, the brain manages to keep the show running smoothly night after night. The flash-lag effect belongs to the relatively small category of temporal illusions. It can be presented in different ways. In one, you watch a computer screen as a black ring crosses the field of view; at some point in its travels (randomly or not doesn't matter, it turns out), the interior of the ring flashes, like this:

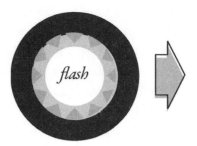

Except that isn't what you see. Instead, and invariably, the flash and the ring do not line up. You see the ring as if it has moved just past the point of the flash, like this:

The effect is so noticeable and so easily replicated that you might think there's something wrong with the computer monitor. But the effect is quite real, a manifestation of the peculiar way the brain processes information. And it is fundamentally baffling: if you are observing the ring at the moment of the flash—that is, if the flash denotes "right now"—how and why does the ring appear *after* right now?

A common explanation, proposed in the nineteen-nineties, is that your visual system is predicting where the ring is going to move. From an evolutionary point of view, this makes some sense. One of the brain's essential tasks is to predict what will happen around you in the near future: precisely when and where the tiger will pounce; where to hold your baseball glove to intercept that fly ball. (The philosopher Daniel Dennett has described the brain as an "anticipation machine.") Likewise, the idea goes, your visual system charts the trajectory and speed of the moving ring, and at the moment of the flash—"right now"—you seem to have caught your brain cheating: it's anticipating just ahead of now (about 80 milliseconds ahead, to be precise) and serving you a picture of where the ring is about to be.

This idea seemed simple enough to test, so Eagleman did. "I believed it was true, about predicting," he said. "I was just curious. But things were not doing what I expected." In standard flash-lag trials, the ring travels along a known path. The prediction hypothesis seems to hold up because, based on the ring's trajectory before the flash, the viewer can accurately anticipate its position after. But, Eagleman thought, what if that expectation is violated? What if, just after the moment of the flash, the ring changed course—veered off at an angle, or reversed, or stopped entirely?

He designed a version of the flash-lag experiment that explored those three outcomes. Presumably, at the moment of the flash, one should still see the ring slightly past the flash point because, based on its preflash motion, that's where you'd predict it was headed. According to the standard explanation, what matters is what happens before the flash; where the ring goes afterward should be irrelevant. But when Eagleman tried his experiment, both on himself and on test subjects, something else happened. In every case, viewers saw the ring positioned along the new trajectory—up, down, reversed—and slightly separated from the flash, even though the change in direction was randomized and unexpected. Viewers seemed to be predicting the expressly unpredictable future with one-hundred-percent accuracy. How could that be?

In one variation of the experiment, the ring began its movement simultaneously with the flash: there was no preflash motion, no way for the brain to anticipate where it would go. Viewers still saw the ring, slightly separated from the flash, along the actual trajectory. In another, the ring moves left to right, the flash occurs, the ring keeps moving in the same direction, and then, some milliseconds *after* the flash, reverses direction. If the reversal occurs within 80 milliseconds after the flash, the viewer sees the ring—and the flash-lag effect—along the new, reversed trajectory, something like this:

Any change in direction—up to 80 msec *after* the flash—influences what people believe they see at the time of the flash. A change in direction immediately after the flash causes the biggest effect; the longer after the flash, the smaller the effect. It seems that the brain continues to gather information about an event such as a flash for as long as 80 milliseconds after the event occurs; this data is fed into the brain's retrospective analysis of where and when that event took place. Eagleman said, "I got myself confused, until I realized there was a simple explanation: it can't be prediction. It has to be postdiction."

Unlike prediction, postdiction is retrospective. Fundamentally, the flash-lag illusion asks where the viewer is in time. The prediction hypothesis presumes, reasonably enough, that because the flash occurs "right now" it must mark "now," the present moment; therefore, the slightly unmoored ring must be an anticipated glimpse of the future, "just ahead of now." In effect, the viewer sits on the flash, gazing forward. Eagleman proposes the converse view. Sure, the flash seems to occur "right now," but the only way you can accurately see the ring *after* now is if you've already arrived there.

It's tempting, then, to think of the ring as the true mark of the immediate present and to think of the flash as marking "just before now" or perhaps even "the beginning of now"—a lingering ghost from the immediate past. But it's even stranger than that, Eagleman contends. Neither the ring nor the flash mark the present; they are both ghosts of the near past. Conscious thought—for instance, the determination of when "right now" is—trails our physical experience very slightly. What we call reality is like one of those live, televised awards shows that has a brief delay built in in case someone curses. "The brain lives just a little bit in the past," Eagleman said. "It collects a lot of information, waits, then it stitches a story together. 'Now' actually happened a little while ago."

We talk about "real time" but we hardly know what that is. Allegedly live television programs insert delays. Telephone conversations obscure the short lag time that arises when communications signals traverse long distances even at light speed. The world's most accurate clocks can only agree on when "now" is by placing it at some agreed-on date next month.

The human brain is in the same bind. In any given millisecond, all manner of information—sight, sound, touch—arrives at different speeds and asks to be processed in the correct temporal order. Tap your finger on the table. Technically, because light outraces sound, the sight of the tap should register some milliseconds before the sound of it. Yet your brain synchronizes the two to make them seem simultaneous. The experience should be even more pronounced when you see someone speak to you from across the room; fortunately it isn't or our days would unwind like a badly dubbed movie. But if you see someone bouncing a basketball, say, or chopping wood more than about a hundred feet away, and you notice carefully, the sound and action will be slightly out of sync. At that distance, the lag between sight and sound is wide enough—about 80 milliseconds—that the brain no longer treats the two inputs as simultaneous.

This phenomenon, known as the temporal binding problem, is a longstanding puzzle in cognitive science. How does the brain track the arrival times of different bits of data, and how does it reintegrate them to provide us with a unified experience? How does it know which properties and events belong together in time? Descartes argued that sensory information converges in the pineal gland, which he envisaged as a kind of stage or theater for consciousness; when stimuli reach the pineal gland, you become aware of them and direct your body to respond. Few people take the idea of a central stage seriously anymore, but its ghost lingers, to the annoyance of philosophers like Dennett. "The brain itself is Headquarters, the place where the ultimate observer is," Dennett has written. "But it is a mistake to believe that the brain has any deeper headquarters, any inner sanctum arrival at which is the necessary or sufficient condition for conscious experience."

Eagleman notes that our brain is made up of many subregions, each with its own architecture and sometimes its own history; it is the patchwork product of evolution over time. The information from a single stimulus—the bright and dark stripes glimpsed on a tiger, say—follows different paths in the brain and suffers different lag times along the way. Neural latency—the time between when a stimulus occurs and when a neuron responds to it—varies greatly depending on the brain region and environmental conditions. The type of data matters too: neurons upstream from the visual cortex, the brain's main unit for processing visual data, respond more quickly and strongly to a bright flash than to a dim one. Picture a wave of horsemen spreading out from a town, calling on horsemen in other towns to carry a message. Some of the horsemen ride faster, others slower. What begins as a single stimulus quickly becomes smeared out in time across the brain.

"Your brain is trying to put together a story of what just happened out there," Eagleman said. "The problem is, we're yoked to this machinery that makes the information arrive at all different times."

One might readily assume that whatever hits the visual cortex first is perceived first. Neural latency is sometimes offered as an explanation for the flash-lag effect: perhaps the brain processes a flash and a moving object at different speeds, and by the time the flash travels from the eye to the thalamus to the visual cortex, the ring has already moved to a new position, so you end up seeing the two in different positions. According to this theory, the timing of events in the brain directly reflects timing in the world. But if this were so, imagine the strange visions you'd see,

Eagleman says. Here's a stack of boxes that differ only in their grade of brightness—dim on the bottom, bright on top:

Now the stack starts moving quickly back and forth across the page. If your brain were "online"—if it perceived the stack strictly in the order in which it processed each box—the bright boxes would register in your awareness slightly before the dim ones did (because a bright stimulus reaches the visual cortex before a dim one), so they would appear slightly ahead in physical space. As a result, you'd see a bowed stack of boxes, as if the dimmed objects lagged behind, like this:

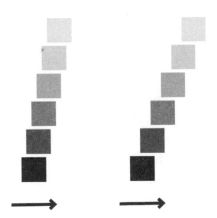

But in fact you see the vertical stack moving. (Eagleman published an experiment that demonstrates it.) For that matter, if your brain were online, you'd see similar motion illusions every time you saw a new vista or image, or turned on the light, or simply blinked your eyes. We don't, which suggests that our perceived timing of events in the real world doesn't directly reflect the temporal order with which those events travel through our neurons. The brain does its processing offline, not online.

Much of the research on the temporal binding problem has focused on the means. How are events bound together in time by the brain? Are they somehow tagged on entry? Is there a timeline or millisecond-scale clock ticking away in the brain's corridors, analogous to what a film editor might rely on, that enables events to be properly synchronized? Eagleman begins with a more straightforward question: *when* is this work done? And it's clear to him that synchrony cannot be achieved on the fly, strictly in the order in which signals arrive. There must be a delay, a buffer period during which the brain collects all the available information from a given moment—which has become smeared out in time within the brain—and serves it up to the conscious mind. In the brain, as in the external world of clocks and universal timescales, it takes time to make time.

In our visual system, the extent of the temporal smear is about 80 milliseconds, or a little less than a tenth of a second, long. If a bright light and a dim bulb flash simultaneously, the dim signal reaches the visual cortex about 80 milliseconds after the bright one. The brain seems to take this interval into consideration. When assessing when and where an event occurred—two simultaneous flashes, say, or a flash within a moving ring—it suspends judgment for 80 milliseconds, to allow the slowest information to arrive. The process of postdiction is a bit like a frame or a net that the brain extends, retrospectively, around an event, to collect all the sensory data that might have been simultaneous to that particular instant. In effect, the brain procrastinates. What we call consciousness—our conscious interpretation of what is happening *right now* (which is about as good a definition of consciousness as any)—is the story that our procrastinating brain narrates to us at least 80 milliseconds later.

It took me ages to wrap my head around postdiction. Time and again I thought I had explained it to myself, only to stop, baffled, for reasons I couldn't quite identify. I'd call Eagleman, and he'd walk me through it again from the beginning, slowly and cheerfully. At last I put my finger on it: if the brain is waiting for the slowest information to arrive—if postdiction is the brain's way of getting the order of events right—why does it still get them wrong in the flash-lag effect? If the brain is waiting to determine what happened "right now," at the flash, why was I not seeing the flash perfectly within the ring? Why was there any illusion at all?

That's where things get strange indeed, Eagleman said. The flash-lag experiment presents the viewer's brain with a question that it rarely asks in the course of daily life: where is this moving object *right now*? Where is the ring at the moment of the flash? As it happens, the brain operates separate systems for judging the positions of stationary things and for tracking objects in motion. When you weave through a crowd at the airport or watch raindrops fall, the brain computes with motion vectors—basically, mathematical arrows of movement—and never stops to ask where a particular person or raindrop is at a particular moment. When an outfielder pursues a pop fly, he does so using the same motion-vector system that a bat uses to catch insects or a dog uses to catch a Frisbee. A frog that had to ask, "Where is the fly *right now*, and *right now*, and *right now*?" would go hungry and, before long, extinct. Many animals, including reptiles, don't even have a positional system; they see only motion. If you stop moving, they can't see you.

"You're always living in the past," Eagleman said. "The deeper issue is, most of what you see, your conscious perception, is computed on a need-to-know basis. You don't see everything, just what's most beneficial for you. It's like if you're driving on the road, your brain isn't continuously asking, 'Where is the red car now? Where is the blue car now?' Instead it asks, 'Can I change lanes? Will I make it through the intersection before that other car crosses?' It's rare that you care about the instantaneous position of a moving object—and until you ask, you don't actually know it. And when you ask, you'll always get it wrong."

The flash-lag effect exposes the gap in the brain's dual approach. In the moments before the flash, you track the ring's motion vector, at no point asking where it is *right now*. The flash prompts the question. It resets the motion vectors; the brain now assumes that the ring's motion began with the flash, at time zero. Before answering the question posed by the flash—where is the ring *right now*, at time zero?—the brain waits 80 milliseconds to gather all the possible visual data from that moment. In the meantime, the ring continues to move, and that additional trickle of information shades the brain's interpretation of where the motion began. As a result, the answer to "Where is the ring right now?" is biased—shifted—slightly in the direction of the ring's movement.

Eagleman devised an experiment to prove it. In the standard flash-lag setup, the viewer sees a single moving ring or dot that passes over a stationary flash. In Eagleman's variation, after the flash, the dot becomes two dots that move away from the flash at forty-five-degree angles. If neural latencies were responsible for the flash-lag illusion, you'd perceive the dot in a position that it actually occupied when its signal struck your visual cortex—along one or the other, or maybe both, of the angled trajectories. But you don't. Invariably, Eagleman's subjects perceived the dot at a point midway between the two, in a position that it never actually occupied. It's as if the two motion vectors had been added up and averaged out—which, Eagleman thinks, is essentially what happens.

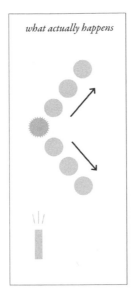
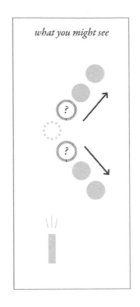
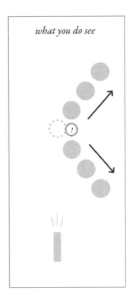

The phenomenon is called motion biasing and it is the key to postdiction. Accept as a given that the conscious mind perceives in retrospect: "right now" already happened. For a short period after that instant, the brain continues to process data (for instance, the dot's motion after the flash) as it settles into a judgment of what happened at that instant. *Where was the dot at the moment of the flash?* The additional motion information biases the final analysis, resulting in an illusion: a moving dot perceived, at the moment of the flash, in a spot it never occupied. Oddly enough, Eagleman's model serves up a result almost identical to the one provided by the predictive hypothesis; in both models, the illusory dot represents the brain's best guess as to where the dot *probably will* appear. Except in fact this judgment is made in retrospect, not in advance—not prediction but postdiction.

Consider again the present. Ask yourself, "What is happening *right now?*" The more narrowly you define the present instant, the more certainly your answer will be a) after the fact, and b) wrong. Just as important, the answer is unknowable—nonexistent—until the moment you inquire. In postdiction, the brain retrospectively extends an 80-millisecond window around an event, to collect all the information that occurred at that instant. But this window is not permanently open, like the open shutter of a film camera. Time in the mind is not a continuous stream of 80-millisecond frames waiting to be reviewed. Rather, the 80-millisecond window is triggered by the question, which is posed only rarely in our daily activities. "You don't have a frame until you need one," Eagleman said. "Then you go collect one."

For thousands of years, philosophers have debated the nature of time. Is it a seamless river or a string of pearl-like moments? Is the present an open frame that glides, stationary, above the flow, or is it just one in a ceaseless series of nows, a single frame on an ending reel of them? Which is correct, the traveling moment hypothesis or the discrete moment hypothesis? Eagleman's answer is, Neither. An event or instant does not present itself to the brain a priori; it is not out there waiting to be noticed. Rather, it comes into being only after it is over, once the brain has paused to address and assemble it. "Now" exists only later—and only because you've stopped to declare it.

One morning I came into Eagleman's lab to try out an experiment that he was still refining; he called it Nine Square. He turned on a computer that the grad students weren't using and sat me down in front of it. Nine large squares appeared on the screen, arranged in a three-by-three grid, like a tic-tac-toe board. One of the squares was a different color than the others, and, at Eagleman's instruction I moved my mouse cursor over it and clicked; the moment I did so the color moved to highlight a different square. I followed with my cursor, clicked that square, and the color moved again. I did it again and the color moved to yet another square. I continued in this way for a couple of minutes, chasing the color around the screen. This was just the warm-up phase, the initial part of any experiment, to help the subject get acquainted with how the setup works, Eagleman said. But in a few moments, he added, and for not much more than a moment, I would have the distinct feeling that time had just gone backward.

Typically when we talk about "the perception of time" we are referring to the perception of duration. How long is this stoplight, and doesn't it seem longer today than usual? How long ago did I put the pasta in the pot of boiling water, and am I about to ruin dinner? But there are other facets of time. One is synchrony, or simultaneity—the sense that two events happened at exactly the same time. Just as important, and often overlooked, is temporal order, which is effectively the opposite of synchrony. Pick two events—a flash and an audible beep, say. If they aren't simultaneous, they must have occurred in succession; how do you perceive which one came first? That's temporal order. Our days are filled with countless temporal-order judgments, the vast majority of which unfold in milliseconds, without conscious thought. Our grasp of causality relies on an ability to assess the proper order of events. You push the button for the elevator and an instant later the door opens—or did the door actually open first? Natural selection has probably played a strong role in shaping the perception of causality. If you're walking through a forest and hear a twig snap, there's an advantage in knowing whether it coincided with your footfall—in which case you probably caused the sound—or whether the twig snapped just before or just after, in which case maybe a tiger did it.

Such assessments are so fundamental that "assessment" seems like too grand a word for it; of course the brain knows what came first and what came second. How could it be otherwise? But Eagleman's experiments with dots and flashes suggested that if the brain can misjudge events that are genuinely simultaneous, perhaps it could mistake the order of things too. "This stuff is shockingly malleable," he said. "We're discovering how plastic the sense of time is." Here's an experiment: you're stationed at a computer monitor and are asked to listen for a beep. Either right before or right after the beep, a small flash occurs on the screen; you're asked which came first, the flash or the beep, and to estimate how much time elapsed between them. You'll have little trouble judging the correct order or estimating the delay, even if the delay is as slight as 20 milliseconds, or one-fiftieth of a second. Now suppose you repeat the exercise but this time there's no beep; instead you push a key on the keypad: you're active rather than passive. Again, either before or just after your keypress, a flash occurs on the screen. If the flash happens first, you'll still be pretty good at estimating how much time elapsed between the flash and your keypress. But if the flash comes after, your estimate will stink. In fact, if the flash occurs as much as 100 milliseconds, or a tenth of a second, after your keypress, it will seem to you as if there was no delay at all; your keypress and the flash will feel simultaneous.

Eagleman designed that experiment with a former student, Chess Stetson, who is now a neuroscientist at Caltech, and they found that, immediately following your action (the keypress), there's a span of about 100 milliseconds in which you can't detect any succession of events; everything seems to happen at once. The critical factor is your involvement. The brain is credit-grabby; it presumes effects for its actions. You act—you merely press a button—and assume that you caused the events that immediately follow. "It's like your brain has a catch-all tractor beam after an action: 'I'll take credit for that as being mine,'" Eagleman says. In the glare of this tractor beam, the true sequence of events—temporal order—dissolves, and a tenth of a second is redefined as no time at all.

By bending time, the brain provides a strange but satisfying service, enhancing our sense of agency and making us seem slightly more powerful than we actually are. In 2002 the neuroscientist Patrick Haggard and his colleagues reached a similar conclusion, with an experiment that asked volunteers to watch the rapidly moving hand of a clock. At her leisure, the subject pressed a key on the keypad and noted when on the clock she'd done so. But sometimes, instead of pressing a key and noting the time, the subjects heard a beep and noted the time: they were passive (listening) rather than active (pressing). And sometimes the conditions were combined and causal: subjects pressed a button, which made a beep 250 milliseconds later, and noted either when they'd pressed the key or when they'd heard the beep (it's impossible to attend to both). Haggard found that when the subject actually caused the beep, the keypress and beep seemed to occur closer together in time than they really did: the keypress seemed to happen a little later (by about 15 milliseconds, on average) and the tone seemed to happen much sooner (by about 40 milliseconds). Causing an event seems to pull the cause and its effect closer together in time, a phenomenon Haggard called "intentional binding."

How does the brain pull off this trick? Most likely, Eagleman figured, it maintains separate expectations about when the keypress will occur and when the beep will occur; it keeps separate timelines and recalibrates them in relation to each other. Calibration is a persistent concern of our brain's daily business. From the constant wash of sensory input, processed at different rates along different neural pathways, it must assemble a coherent picture of events and actions, causes and effects. Working backward from the signals, it must determine which stimuli occurred first and which were simultaneous, which were linked and which were not. When you catch a tennis ball, the sight of the ball striking your hand reaches your brain faster than the tactile sensation from your palm, yet somehow you experience the two data streams as simultaneous. Or, to see it from the other end, your brain receives two data packets, one tactile and the other visual, separated by some milliseconds of delay; how does it know they belong to the same event?

Moreover, the speed of the sensory signals can vary depending on conditions, so the brain must be able to shift its assumptions about when the source event occurred. Say you're tossing a tennis ball outdoors and then go indoors, into a dim room. Your neurons process dim light less quickly than bright light; indoors, the visual input from your activities arrives more slowly than it did when you were outside. Your motor actions must take the shift in timing into account or you'll toss and catch like an awkward teenager. Fortunately, your brain recalibrates; it calls the new timing "normal" and shifts its other sensory expectations accordingly. It recalibrates constantly throughout your day, working to serve up a smooth interpretation of reality as you switch activities, move from environment to environment, and speed up and slow down.

When an action (the press of a button) and its effect (a flash) seem to move closer in time, or when the delay disappears entirely, what you're experiencing is a recalibration, Eagleman posits. In general, your brain expects your motor actions to produce their intended effects immediately, without delay. So when it recognizes an event that was caused by something you did—or more to the point, an event that follows your action within a tenth of a second—it recalibrates and gives the event the same time stamp as your action: time zero. Cause and effect are made simultaneous. A tenth of a second is a slight amount of time but it's not negligible and is certainly enough, in other situations, to register in one's awareness. Clearly, there are moments when the brain concludes that consciousness is not in its—or our—best interest.

This illusion predicts an even stranger one. If, by recalibrating, your brain can make cause and effect seem to occur simultaneously, perhaps it can be fooled into altering the temporal order still further, to make the effect seem to come before the cause. With Stetson and two other colleagues, Eagleman designed an experiment to test the idea. Once again, volunteers pressed a button to make a light flash, but Eagleman inserted a delay of 200 milliseconds—a fifth of a second—between the keypress and the flash. The subjects adapted to the delay almost immediately and didn't notice it, provided the time lag didn't last much more than 250 milliseconds; as far as viewers were concerned, the keypress and the flash occurred simultaneously. Your brain performs causal sleights of hand all the time in daily life. Whenever you type a letter on a computer keyboard, for instance, roughly 35 milliseconds elapse before you see that letter on your screen—a delay that goes unnoticed by you. (Eagleman actually measured that delay while designing his reverse-causality experiment, in order to factor it out.)

Once his subjects adjusted to the delay, it was removed; the flash occurred exactly at the moment the key was pressed. On those occasions, something peculiar happened: the volunteers reported that the flash occurred *before* they'd pressed the button. Their brains had recalibrated to set the delayed flash alongside the keypress, at time zero. Under that redefinition, a flash that then occurred sooner than the (expected) delayed flash would be perceived as occurring before time zero and therefore seemed to occur before the keypress. Cause and effect—time, or at least temporal order—seemed to reverse.

Eagleman had since refined the experiment into a quicker format, the Nine Square version that I was trying out. I clicked again on the square that changed color, saw the color move to another square, then clicked on that one. I knew in advance that there was a 100-millisecond delay between my mouse click and the movement of my cursor, but I didn't notice it; the fact of my click—through which my brain declared itself the author of whatever happened next—rendered the subsequent delay transparent. So I didn't notice when, after a dozen or so clicks and jumps, the delay was removed. But I did notice the result: to my amazement, just before I clicked the mouse, the colored square jumped to its next position, to precisely the square where I'd planned to send it.

It was unsettling, to say the least. The computer had seemingly guessed my next move and made it for me. I ran through the test a few more times just to make sure that what I thought had happened really did happen, and each time it did: as I prepared to move the cursor, the colored square moved on its own, to exactly where I meant to send it. I knew it would happen but it happened anyway, again and again. The experience was so distinct, the repositioning of the colored square so obviously divorced from my own impending mouse click, that as soon as I noticed the movement, I found myself trying to stop my finger from clicking the button, which of course was causally impossible: the square had already moved on, which meant I'd already moved it, which meant I was trying to prevent something that I'd already done. And because I couldn't help it, and because I'd already done it, I pushed the button. Up until then I had enjoyed Eagleman's research as one might enjoy a series of rides at a carnival; with this one, it was as though I'd suddenly dropped through a crack into another dimension.

Once, after giving a talk at a college about the phenomenon, Eagleman was approached separately by two audience members who described a curious experience. The campus had just installed a new phone system, they told him, and it behaved strangely: you'd dial your party and the phone would start ringing at the other end before you'd hit the last digit. How could that be? Eagleman suspects that the illusion arises because the individual has switched from using the computer keyboard, with its 35-millisecond delay, to the phone keypad, which has a shorter delay between each keypress and its effect. Your

brain, having calibrated to the computer delay, applies its sense of simultaneity to the phone—and finds itself surprised by the immediacy
of its actions.

The illusory reversal of causality is disconcerting but it's the result of an
entirely normal, and highly adaptive, aspect of our perceptual experience. The only way to get temporal order right and to correctly discern
causes and effects, when sensory data is flooding in at different speeds
along different neural pathways, is by constantly recalibrating. And the
quickest way to calibrate the timing of incoming signals is to interact
with the world. In causing an event, you render its outcome predictable: the effect should immediately follow your action. You've imposed
a definition of simultaneity on your sensory experience—a baseline, or
time zero, against which you can estimate the temporal order of other,
related data. "Every time you kick or knock something, the brain assumes that whatever happens next is simultaneous," Eagleman said.
"You're forcing simultaneity on the world." To act is to expect, and to
expect is to time.

This perspective has given rise to what he concedes is one of his
more out-there theories. Recall that the brain is saddled with various
lags and latencies: a bright flash registers in the neurons more quickly
than a dim flash from the same source; red light registers before green
light, and both register before blue light. If you glance at an image or
scene that contains red, green, and blue wavelengths—an American flag
spread out on a lawn—its portrayal in your brain will be smeared out
slightly in time, and the scale of that smear can vary further depending
on whether you're standing in the shadows or in full sun. Yet somehow
your brain registers the data streams as having a single, simultaneous
origin. How does a downstream neuron know which input came first or
that the three colors belong together? How does the system learn that
red always arrives before green, and that green arrives before blue, and
that the arrival of a "red-then-green-then-blue" signal means that the
inputs originated simultaneously from a single event? Otherwise, you'd
see the flag as a stream of emerging colors: first the red stripes, then the

blue field of stars, then the lawn behind it. Your visual experience would be one big, psychedelic swirl.

To unify this experience, the brain needs a way to intermittently recalibrate the visual streams and intermittently set the time to zero. Eagleman thinks that we may achieve this by blinking. Blinking has the obvious benefit of keeping the eyes moist. But it also has the effect of flicking the brain's lights off and back on again. In the moment when the lights first return, your senses might experience a red-green-blue blur. But after much repetition—thousands of times every day—the brain learns that a red-green-blue blur spanning some tens of milliseconds is equivalent to simultaneity. We think of blinking as a passive act but it may also serve as an active one, as willful as pushing a button and a way of exerting one's intention on the visual world. It's a training mechanism for the senses, a forced reboot; simultaneity occurs not because you've received events as simultaneous but because you've made them so with your eyes. The blink says, "I name this 'now,'" and your actions and perceptions that follow shortly thereafter reorganize themselves around the declaration: *This is now. This is now. This is now.*

Once I was invited to give a talk in Italy, as part of a panel discussion. I was slated to speak last, so I spent the afternoon listening to my fellow panelists, all of whom were, and spoke in, Italian, a language I don't know. Their words swirled around me; from time to time, when it seemed that something funny or insightful had been said, I nodded my head appreciatively, as if I understood. I felt like Pluto at the dark edge of the solar system, observing the spark of a distant Sun and thinking how much more pleasant it would be to live among the inner planets.

After the fourth or fifth speaker I noticed a set of headphones on the table in front of me. The proceedings were being simultaneously translated from Italian into English, and vice versa, courtesy of someone in a glass booth that I suddenly noticed in the back corner. The translation helped, a little; with the headphones on, I could understand that the current speaker, an academic philosopher, was somehow linking Charles Darwin to Newtonian physics. He was rambling, or it was over my head, or both, and the translation felt off. There were long pauses in which I could all but hear the translator, a young woman, laboring to make sense of it. I peered toward the booth and saw two figures inside. Before long the woman's voice in my headphones gave way to a young man's, which more quickly and articulately translated the Italian into English.

When my turn finally came, two or three people in the audience put on their headphones, which left me with a sinking feeling about everyone else. I apologized for not speaking Italian and then launched into my talk, but I spoke slowly, with the vague notion that this would help the translator. I soon realized my mistake: by speaking at half my normal speed, I effectively had given myself half as much time to cover my forty-minute talk. I tried to edit on the fly—I skipped examples, dropped segues, lopped off whole branches of thought. The result was a talk that made less and less sense to me even as I heard myself speak it. The faces of the people with headphones were as blank as those of the people without them.

In 1963, the French psychologist Paul Fraisse published *The Psychology of Time*, a review of the previous century or so of temporal research and the first book to address the field as a whole. It examined every facet of time perception, from temporal order to the apparent length of the subjective present, which, after considering the numerous studies, Fraisse defined as "the time necessary to pronounce a sentence of 20 to 25 syllables"—perhaps five seconds at most. My own present could not have been shorter or felt longer. Fraisse argued in addition that many of our feelings about and perceptions of time "have their origin in the consciousness of frustration caused by time. Time either imposes a delay on the satisfaction of our present desires or it obliges us to foresee the end of our present happiness. The feeling of duration thus arises from a comparison of what is with what will be." Boredom in particular is "the feeling that results from the non-coincidence of two durations"—the duration you're stuck in and the duration you'd prefer to inhabit. It's another version of Augustine's "tension of consciousness," and, as I spoke, I was all too conscious of the tension in my bored listeners. I should have felt like the Sun radiating knowledge onto the audience. But I was still Pluto, with the telescopes of the inner planets trained on me and wondering what to make of this distant, foreign, frozen object.

That evening, at a dinner for the panelists, I met my translator, whose name was Alphonse. He was a graduate student in linguistics, fluent in French and Portuguese as well as English; tall and thin, with dark hair and round glasses, he brought to mind an Italian Harry Potter.

We agreed that "simultaneous translation" is plainly an oxymoron. The rules of syntax and word order differ among languages, so one can't translate strictly word by word from one language to the next. Always the translator is holding back a little something from the listener: he hears what sounds like a key word or phrase and must keep it in mind until something later in the speaker's sentence gives it meaning and enables him to start translating aloud, even as the speaker forges on with new words and insights. If the translator waits too long, however, he risks forgetting the original phrase or losing the sense of the ongoing stream. "Simultaneous" implies an activity that is set purely in the present; really it is a continuous expression of memory, presented in such a way as to seem transparent.

The challenge is more acute when translating between languages that belong to different families, Alphonse said; German to French, say, is more difficult than Italian to French or German to Latin. In German and Latin, the verb typically comes toward the end of the sentence, so the translator often must wait to hear the conclusion of the sentence before the beginning makes enough sense to translate. If he is translating into French, where the listener expects a verb at the beginning of the sentence, the translator can either wait the extra time or guess where the sentence is headed.

I told Alphonse that I often confront a similar problem dealing purely in English. For a long time I used a tape recorder when I conducted interviews, in order to catch every word. But what I gained in accuracy I lost in time: an hour-long interview might take four hours to transcribe, for what might amount to just a few useful insights or quotes. Taking notes by hand was hardly more practical: my handwriting is awful, worse when hurried. Sometimes, when on the phone, I can type on my computer while my subject talks; that makes things neater, at least. But I type more slowly than most people talk. Often enough I'll look back through my notes and come across a meaningless fragment such as this one:

"If something surprising to it, have a faster."

In this case I was lucky: I had corrected the note shortly after I made it, so I remembered what the speaker actually said: "If something surprising happens, you have a faster reaction to it." Looking over the fragments I can see what went wrong. I began on solid footing, managing to get the first three words—"If something surprising"—correctly. But my subject spoke too fast and I dropped the thread. So I decided to try to remember the key aspect of what he was saying right then (the verb "happens") and jot it down as soon as he paused. Like a juggler, I would toss his words into the near future (that is, into short-term memory) and catch them a moment later. In the meantime I wrote down the next words I heard ("to it"), and as he continued to talk—without pause, alas—I typed the few words ("have a faster") that I could still recall from a moment earlier. All this without conscious thought, in the time it takes to speak a short sentence, repeated countless times over the course of an hour-long conversation. It's a wonder that I managed to retain any information at all. (I might have fared better if I'd followed Alphonse's approach, to write down the key phrases, not try to remember them.)

In Alphonse's telling, a translator resembled a person that Augustine might have recognized—stretched taut between the past and the future, between memory and anticipation. Alphonse estimated that the average translator can absorb a lag of anywhere from fifteen seconds to a minute between what she hears and her own "simultaneous" translation of it. The better the translator, the longer the lag—the more information she can hold in her head before producing a translation. A translator may prepare for three or four days in advance, to get a sense of the jargon she will encounter. Alphonse said that when it's going well, live translation is a bit like surfing.

"You must spend the least possible amount of time thinking about words," he said. "You are trying to go with the flow, you are listening to a rhythm. You don't want to stop. Otherwise you fall behind, you are losing time, you will be lost."

Consider a sentence that starts *here*, continues for a few words, wanders off on a clause or two, and then ends *here*. It took me a few seconds to compose that sentence, and it may have taken me several years to get around to committing it to paper. But you read it in perhaps two seconds—quickly enough that you barely registered that you read it or that any time was needed to read it. By some measures, that is the present.

But of course it isn't, technically. A great deal of cognitive activity unfolds in that span of time, although the brain—or the mind, it's not always clear which—goes to great lengths to disguise it from one's conscious self. As you read, and without your really noticing, your eyes flit around the page to preview upcoming words or review past ones. Studies indicate that as much as thirty percent of your reading time is spent going back over words you've already read. If you eliminate these "regressions," perhaps by using an index card to cover preceding lines of text, you can substantially increase how fast you read, at least if one believes the claims made by certain speed-reading courses.

In his book *Mindworks: Time and Conscious Experience*, the German psychologist and neuroscientist Ernst Pöppel describes an experiment he conducted on himself in order to reveal how discontinuous the experience of reading actually is. He picked out a brief passage from an essay by Sigmund Freud on the unconscious mind:

> SOME REMARKS
> ON THE CONCEPT OF THE UNCONSCIOUS
> IN PSYCHOANALYSIS
>
> I should like to present in a few words and as clearly as possible the sense in which the term "unconscious" is used in psychoanalysis, and only in psychoanalysis.
> 1. ▶ A thought—or any other psychic
> 2. ▶ component—can be present now in my con-
> 3. ▶ scious, and can disappear from it in the next instant; it can, after some interval, reappear completely unaltered, and can in fact do so from

As he read, a device tracked the movement of his eyes on the page, recording where he looked and for how long. He then drew a rough graph of the motion: the graph moves upward as he reads across the first line of text, left to right, and sweeps back down when he reaches the end of the line and starts the second one. Although his experience of reading is smooth, his eye movements clearly are not. His visual path resembles a series of steps, his eyes halting for two- or three-tenths of a second to absorb meaning, then skipping ahead to the next point of absorption.

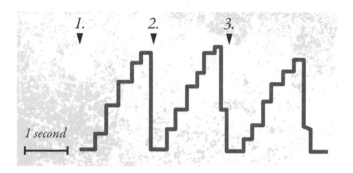

Next Pöppel read something slightly more challenging, a passage of roughly equal length from Immanuel Kant's *Critique of Pure Reason*. The added difficulty of the text is made evident by the time code: compared to the Freud, each line from Kant took Pöppel about twice as long to read, and his eyes paused to absorb information about twice as often.

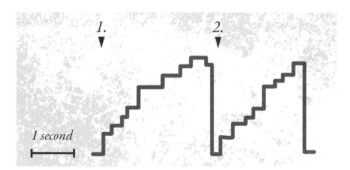

Finally Pöppel graphed his effort to read Chinese—a language, he noted, of which "the author is unfortunately ignorant." One can see that he spends several seconds attempting to read one or two characters, but two-thirds of the way through the first line he gives up and skips to the end:

"reading" abandoned

next line

1 second

Pöppel's point is that what we sometimes experience as "now" is in fact bursting with cognitive activities—syllables, eye saccades, the harvesting of meaning—that can't be accessed entirely through introspection. Moreover, he notes, the busyness within each moment is carefully orchestrated; the speaking of successive syllables and the movement of the eyes from written word to written word is synchronized "like the cars of a train according to a timetable." But how?

In 1951 the Harvard psychologist Karl Lashley addressed the relationship between time and language in a now-classic paper, "The Problem of Serial Order in Behavior." Lashley noted that, for words to convey meaning, they must be presented in a particular order. The phrase "Little a mary had lamb" means nothing, but rearranged—"Mary had a little lamb"—the statement is comprehensible. As my Italian translator noted, the rules of syntax vary from language to language; in English, for instance, an adjective typically precedes the noun it modifies ("yellow jersey"), whereas in French it comes second (*"maillot jaune"*). These rules are malleable and socially acquired, and they change over time. Nonetheless, in any given language, the order of words is significant: it signifies.

For the most part we give syntax no conscious thought; it unfolds seemingly by itself, in a timescale just below awareness. (Your brain is so eager to find order that when you first scanned the words "Little a mary had lamb," you may have seen right through to its intended meaning and not noticed the scramble.) And sometimes we get the order wrong. Lashley noted that when he types, he sometimes misplaces letters, writing "thses" instead of "these" or "wrapid riting" instead of "rapid writing." (As a matter of fact, when I typed the previous sentence, I accidentally wrote "dypet" for "typed," before correcting it.) What's notable about these mistakes is that they are often errors of anticipation: a letter or word that should arrive later instead appears sooner—now—as if the mind's eye (for want of a better term) had wandered ahead, distracting the fingers from their current task. How do we generate the correct temporal order without thinking about it? Lashley considered this problem "both the most important and also neglected problem of cerebral psychology."

According to Pöppel, the organizing mechanism that Lashley had in mind, although Lashley did not use the word expressly, is a clock. "This construction or stringing together of the words into their proper places occurs—under guidance of a mental plan—by means of a clock," Pöppel writes. "The clock in the brain ensures that all administrative functions, all regions of the brain involved in putting the train of words together are running synchronously, so that, in relation to the overall plan, they are able to discharge their appointed tasks at the correct time." This cerebral clock is "the precondition for expressing a thought through the medium of properly ordered words." Without it, we couldn't make ourselves known.

"All complex behaviors involve time," Dean Buonomano, a neuroscientist at the University of California at Los Angeles, told me shortly after we first met. "We can't fully understand how the brain navigates the world without understanding the temporal component."

Buonomano is among the few researchers who studies physiological time in the millisecond range; he has coauthored papers with many other scientists in the field, including David Eagleman, and is persistently interested in understanding how the time we experience in our daily lives relates to, and emerges from, the activity of our neurons. Neuroscience is a young field, he said, and is adept at answering certain kinds of puzzles, such as how the brain interprets spatial information. For instance, you're able to distinguish between a vertical line and a nearly vertical line thanks to individual neurons in your cortex, discovered in the nineteen-sixties, that respond separately to different orientations. Points in space map to arrangements of retinal neurons somewhat in the way that musical notes map to keys on a piano. But ask neuroscientists how you can tell that one line remained onscreen longer than the other, and they'll likely be stumped.

"I think time has been neglected because the science isn't mature enough to cope with it in a sophisticated manner," Buonomano said. Even mentioning the word *time* can prompt a series of definitions and qualifications. "That's the funny thing about this field," he said. "Nobody can quite describe what we're talking about."

Buonomano and I met in the lobby of a café, then walked across campus, down an allée of palm trees, toward his office. His fascination with time began at age eight, when his grandfather, a physicist, gave him a stopwatch for his birthday. He used the watch obsessively to time his performance on various tasks that he assigned to himself, such as completing a puzzle or walking a city block. When he published a paper in the journal *Neuron* laying out his own research on how time works in the brain, the cover image was a photograph of his stopwatch.

Buonomano noted that, when thinking about time on the millisecond scale, it's important to distinguish between temporal order and timing. Temporal order is the sequence in which events occur over time; timing is duration, how long an event lasts. The two phenomena are distinct but they work together in subtle ways. The simplest example is Morse code. Developed in the eighteen-thirties and -forties, for use with the telegraph, Morse code is a language composed entirely of pulses—dots and dashes—and the silences between them. Modern international Morse code has five linguistic elements: a basic dot, or "dit"; a dash, or "dah," equal in length to three dits; a silent gap, one dit long, between the dots and dashes within one letter; a silent gap, three dits long, between letters; and a seven-dit-long silent gap between words.

146

To correctly express and interpret Morse, you have to get both the temporal order and the timing right. Reversing the sequence of

• • • • — results in — • • • •
(the number 4)   (the number 6)

and mistiming the duration of the middle element of

— • • instead produces — — •
(the letter D)   (the letter G)

A good coder can generate and translate forty words per minute; the record is more than two hundred words per minute. At those rates, a single dit can last from thirty milliseconds, or three-hundredths of a second, to as little as six milliseconds, or six-thousandths of a second. The *Wall Street Journal* once interviewed Chuck Adams, a retired astrophysicist and coder who spent his free time translating novels into Morse code. After the release of his hundred-word-a-minute version of H. G. Wells's *The War of the Worlds*, Adams received an email from a man who complained that he had made the interval between words ever so slightly too long—eight dits instead of the standard seven. The man was vexed by a span of time that, at Adams's rate of translation, was only twelve-thousandths of a second long.

To accurately discern such brief durations—not just once but hundreds or thousands of times per second—demands a keen sense of duration. Pöppel was right: language requires a clock. But where does this clock reside, and how does it work? Buonomano cautioned against thinking about the millisecond clock in literal terms; the kinds of models commonly devised to describe the mind often fail to capture the complexities of actual neurons at work. The standard explanation for how we estimate duration invokes what is often called the pacemaker-accumulator model of time, which supposes that somewhere in your brain is something like a clock, maybe a set of neurons that oscillate at a steady pace; these generate pulses or ticks that are somehow collected and stored. A certain number of ticks adds up to, say, ninety seconds—and when their sum accumulates much past that amount, you become aware that the red light you're sitting at seems to be taking too long.

But the realm of neurons isn't so tidy. "It's very hard to apply the metaphor of real life to the brain life," Buonomano said. He also believes that the millisecond clock is not something as distinct as a particular group of brain cells. Far more likely, millisecond timekeeping is a process that is distributed across the network of neurons, located in no one place in particular. "Timing is such a fundamental aspect of processing, it's not going to be assigned a master clock for timing," he said. "You don't need a master clock; that's not a robust design."

When a stimulus reaches the brain—for instance, when a dit of Morse code strikes the auditory nerve or a flash registers with the eye—it prompts a relay of electrical excitement among the neurons. A signal passes from one neuron to the next across a small gap, or synapse, by means of neurochemicals; these induce the second cell to fire and transmit its own electrical signal. Picture a scientist tossing a door key across the corridor to a colleague. But it takes a little time, maybe ten to twenty milliseconds, for that second neuron to fire up and also to recover. If another signal comes along within that window of time, it will encounter the neuron in a different state of excitement than the previous signal did. Karl Lashley wrote that the situation might be best illustrated "by picturing the brain as the surface of a lake." A stimulus occurs, generating a signal that enters the network of neurons and produces ripples of excitement, like a stone dropping into water. Another signal follows and adds its own pattern of ripples to the already rippling surface, and so on. It's like that throughout the brain all the time. Neurons are not sitting idle, waiting for a dit of Morse code to jolt them into action; they are constantly busy—relaying information, briefly resting, relaying again. "The input is never into a quiescent or static system, but always into a system which is already actively excited and organized," Lashley wrote.

These ripples are ephemeral, lasting a few hundred milliseconds at most, Buonomano said. Nonetheless, it means there's a brief span in which the network retains information about what just happened. Two states of the network exist at once: the pattern of activity from the most recent stimulus and the slightly different, short-lived residue of the previous one, what Buonomano calls the "hidden state." It's a transient kind of memory, and it's the essence of the millisecond clock. The juxtaposition of the two states—the successive presence, absence, or number of excitatory spikes from a subset of neurons—reveals information about how much time has passed between the two. The clock isn't a counter so much as a pattern detector, comparing subsequent snapshots of the ripples on a pond and converting their spatial information into temporal information: state A and state G, superimposed, suggest that 100 milliseconds has transpired; state D and state Q together suggest 500 milliseconds, and so on. On a computer, Buonomano ran simulations of networks of neurons that incorporate these hidden states and found that his model works—the networks can discriminate different intervals of time.

The model also leads to an important prediction, Buonomano said. If two stimuli, such as two identical audio tones, follow one another by 100 milliseconds—more quickly than the network can reset—the second one will enter a network still rippling with the excitation of the first one; moreover, that hidden state will alter how the new state plays out. "The current output of a neuron is dependent on what happened in its recent past," he said. In other words, if identical stimuli follow in close succession, you'll experience them as having different durations. Buonomano designed a clever series of experiments to demonstrate it. In one version, volunteers listened to two brief tones in quick succession; then they were asked to estimate the duration of the interval between them. The interval varied, and discriminating among them was easy enough until Buonomano introduced a "distractor" tone, equal in duration and frequency, right before the target pair. If it preceded them by less then 100 milliseconds, listeners became far less accurate at estimating the correct duration of the interval between the two tones.

What's actually happening, Buonomano said, is that the distractor alters the perceived duration of the first tone, which throws off the estimate of the interval between the pair. In a variation of the experiment, volunteers listened to a pair of tones, one longer than the other, in quick succession, and were then asked which of the two came first. If a distractor tone preceded the pair by 100 milliseconds, the estimates were much less accurate—the listener had trouble determining which tone was longer and, as a result, the correct sequence in which they occurred. At the millisecond scale, timing and temporal order are entwined. Indeed, other researchers have found that people with certain forms of dyslexia have trouble designating the correct order of two phonemes if they follow in quick succession; it's possible that this results from an inability to correctly estimate durations and intervals on the millisecond scale. In any event, Buonomano's model suggests that although there may be a millisecond clock in the brain, it's a clock that neither ticks nor counts them.

I n 1892, William James, fifty years old and "disliking laboratory
work," he wrote, handed over the directorship of his Harvard psychology lab to Hugo Münsterberg, a German experimental psychologist whom he had befriended three years earlier in Paris, at the
First International Congress of Psychologists. Münsterberg had studied
in Leipzig under Wilhelm Wundt, James's mentor, and is regarded by
many historians as the first to embrace the application of psychology
to industry and advertising. He developed psychological tests to help
the Pennsylvania Railroad and the Boston Elevated Railway Company
hire the safest engineers and streetcar drivers, and, after some study, he
suggested that one way to increase worker productivity was to rearrange
the office to make it harder for employees to talk to one another on the
job. He was the author of numerous books, including *Business Psychology* and *Psychology and Industrial Efficiency*, as well as of popular articles
such as "Finding a Life Work," published in 1910 in *McClure's Magazine*, which proposed that psychological experiments could help reveal
"a man's true calling" and so counter "the reckless choosing of careers
in America."

Münsterberg is also widely recognized as the first film critic. He was captivated by early cinema and in such essays as "Why We Go to the Movies" and in his 1916 book *The Photoplay: A Psychological Study*, he argued that film should be considered a form of art in part because its effects so closely mirror the workings of the human mind. Münsterberg incorporated the medium into his own work, developing a series of psychological tests that could be administered to theater audiences before the main feature to help people "learn what characteristics equip one for special kinds of work, so that each individual may find his proper setting," he said in a speech at the First National Motion Picture Exposition, in 1916. One test, intended "for the executive type of mind, which must be able to grasp the meaning of a situation as soon as it presents itself," showed viewers a series of scrambled letters and asked them to build names out of letters that were arranged in a different order from the original one.

One of the Harvard professor's "ideographs" or visual psychology tests appearing in "Paramount Pictographs," a screen magazine shown at the Stanley. The letters in the jumble to the left are first thrown on the screen. Several seconds elapse. If you can unspe'l and respell them into the word Washington, you are blessed with creative ability.

The advent of cinema introduced looser possibilities for narrative, the historian Stephen Kern has written. Whereas photography arrested time, film liberated it; a story could jump forward, back, or sideways, at every speed. Reverse the projector and time could reverse too: a man could leap feet-first from the water and land safely on the shore; scrambled eggs could revert to yolks. In *The Culture of Time and Space, 1880–1918*, Kern quotes Virginia Woolf: "This appalling narrative business of the realist: getting on from lunch to dinner, it is false, unreal, merely conventional." For Münsterberg, the ability of cinema to skip forward and back in time was a near-perfect simulation of the workings of human memory; the close-up shot emulated the intimate perspective of a focused observer. "The camera can do what in our mind our attention is doing," he wrote. Elsewhere he added, "The inner mind which the camera exhibits must lie in those actions of the camera itself by which space and time are overcome and attention, memory, imagination, and emotion are impressed on the bodily world."

In the decades since, film and video have become the primary metaphor offered to explain, in popular terms, how the brain perceives time. The eye is our camera, our lens; the present is a snapshot of some brief, perhaps even measurable duration; and the passage of time is a stream of such images. Your memory tags these single frames as they are recorded, enabling events and stimuli to be reassembled and recalled later, movie-like, in their correct order. This view of time has settled deep into the grain of neuroscience, and much of David Eagleman's research is aimed at dispelling it. Time in the brain, he would like us to know, is not like time in the movies.

One afternoon in his office, he was eager to tell me about a paper he'd recently written on an illusion known as the wagon-wheel effect. The illusion can often be seen in old westerns, when the spoked wheel of a moving stagecoach appears to be rotating in reverse. The effect results from a mismatch between the wheel's rotation rate and the frame rate of the camera that's filming it. If a wheel spoke travels more than one-half but less than a full rotation between movie stills, the spokes will look like they're rotating backward.

The illusion can arise in real life under the right lighting conditions. Perhaps, during a long meeting in a conference room, you've looked up at the ceiling fan and watched as it seemed to turn in reverse. The immediate cause is the fluorescent lights; flickering at a rate just beyond conscious notice, they create a subtle stroboscopic effect, breaking the continuous motion of the fan into a string of discrete images that are rapidly flashed on your retina, just as a film projector rapidly flashes still images on a screen. The mismatch between the fan's rotation rate and the flicker rate of the lights produces the illusion.

In rare circumstances the illusion can be seen under continuous sunlight. In 1996 Dale Purves, a neuroscientist at Duke University, managed to re-create the phenomenon in the laboratory. He painted dots around the perimeter of a small drum and had subjects watch from the side as it rotated rapidly. As the drum turned to the left, the dots moved to the left—and then, after a bit, the dots seemed to reverse and start moving to the right. Not all of the subjects saw it; some saw it after several seconds had passed; for others it took minutes. And there was no predictable rotation rate at which the reverse suddenly occurred. Nonetheless, it was seen; it happened.

Why? Purves and his colleagues argued that their laboratory illusion, like the wagon-wheel illusion, was evidence that our visual system is clocked like a movie camera: the illusion arises from a mismatch between our perceptual frame rate and the rotation rate of the drum. That the illusion should arise under constant lighting "suggests that we normally see motion, as in movies, by processing a series of visual episodes." Several other scientists cited the study as evidence that we process the world as a series of discrete perceptual moments.

Eagleman was skeptical; if we really do perceive in discrete moments, akin to movie frames, then the results should be more predictable and regular. For instance, the illusory reversal ought to occur reliably when you rotate the wheel at a certain rate. As counterevidence, he conducted what he calls "my fifteen-dollar experiment." At a junk shop he bought a mirror and an old record player. He made a series of dots on a small drum, which he placed on the turntable, to replicate the original experiment; then he placed the contraption in front of the mirror. Now his subjects could simultaneously watch the real drum rotate to the left and its reflection rotate to the right. If the brain perceives in discrete snapshots, like a movie camera, then both drums, real and mirrored, should reverse direction at the same time.

That didn't happen; both drums were seen to reverse direction but not simultaneously. Eagleman concluded that the illusion doesn't involve any sort of perceptual frame rate or have anything to do with how we perceive time. Rather it's related to the waterfall illusion, or motion aftereffect, and involves a phenomenon called rivalry. As you watch the dotted drum rotate right to left, a large group of neurons that detect leftward motion are stimulated. But due to a quirk of how motion detection operates, a smaller number of neurons that detect rightward notion are stimulated too. The result is something like an election. Most of the time the majority wins and you perceive the drum's motion correctly. But statistically the minority perception has a very small chance of prevailing, and very infrequently it does, generating the illusion of backward motion. "It's competing populations of neurons," Eagleman said. "Every once in a while the little guy gets to win."

The movie-camera metaphor persists in neuroscience in subtler guises. Imagine that a series of identical images—of a shoe, say—flashes rapidly on a screen in front of you. Although all the images are equal in duration, the first one will always seem to last longer than those that follow—as much as fifty percent longer, in controlled studies. This is known as the cameo, or debut, effect. (The effect also occurs, though less markedly, with auditory tones, such as beeps, and with tactile pulses.) Likewise, if the series of identical images is interrupted by a novel one—if, say, a boat appears somewhere in the sequence of shoes—it too will seem to last longer, even though it has the same duration as the others. Scientists call this the oddball effect.

The standard explanation invokes the pacemaker-accumulator model of time: somewhere in your brain, operating over small time-scales, is something like a clock; it steadily generates pulses or ticks, which are gathered and saved. Now along comes an oddball image. Being novel, it draws your attention, which increases the rate at which you process data about the oddball, and this causes your internal clock to tick slightly faster while you observe it. Because your brain collects comparatively more "ticks" while viewing the oddball, you perceive the oddball's duration as longer. It's as if you were watching a movie and the appearance of the oddball image caused the frame rate to momentarily slow, stretching out the moment. One scientist has referred to the oddball experience as a "subjective expansion of time."

That didn't sound right to Eagleman. Imagine you're watching a chase scene in a movie and the police car flies off a ramp. If you slowed that sequence, both the audio and the video would be affected, so you'd also hear the siren at a lower pitch. In real life, though, duration distortions don't seem to involve more than one sensory modality at a time. "Time is not one thing," Eagleman told me: time in the brain is not a unified phenomenon. So what accounts for the oddball effect? Probably not attention, he thinks. For one thing, attention is slow. When you suddenly "pay attention" to something, it takes at least 120 milliseconds—more than a tenth of a second—for your attentional resources to zero in on the target. Yet you'll experience the oddball effect even when images are flashed at you at a much faster rate. Moreover, if attention is what makes time dilate, then even more attention-grabbing images should make the oddball effect even more pronounced. But when Eagleman ran the experiment with "scary oddballs"—pictures of spiders, sharks, snakes, and other items from an international database of images ranked by their emotional salience—they didn't slow down time any more than regular oddballs do.

Maybe the standard explanation had it all backward, he thought. It's not that debut and oddball images seem to last slightly longer than normal; their durations *are* normal. Rather, the durations of the subsequent images, which by now are familiar to the brain, are slightly less than normal, so the debut and oddball images feel long in comparison. The oddball doesn't dilate in time: the familiar ones contract. Studies of the brain's physiology suggest that something like this is at work. Using electroencephalograms, PET scans, and similar methods of monitoring the activity of neurons, scientists have found that as subjects watch (or hear, or feel) a sequence of repeated or familiar stimuli, the firing rates of the relevant neurons diminish as the series progresses, although the viewer doesn't consciously experience any change. It's as if, with each successive viewing of an identical image, the neurons become more efficient at processing it. The phenomenon, known as repetition suppression, may be a way for the brain to conserve energy; it may also be a way of enabling the viewer to react more quickly to repeated or familiar events. The neurons basically ease up, an economizing of which the conscious mind is mostly unaware.

That might explain the debut and oddball effects. In the standard explanation, the oddball image draws more attention, which requires additional energy, which seems to dilate the oddball's duration. But if repetition suppression is responsible then the opposite is happening: the successive images contract in duration, the oddball seems to expand by comparison, and attention follows. Attention doesn't distort time; time distorts in order to draw your attention. It's one more blow to the ego; we think of attention as the expression of our conscious self—"I'll look at this now"—but it's yet another prompted response, like the audience laughter in those allegedly live-before-a-studio-audience sitcoms.

It's commonly assumed that temporal illusions arise because at some level the brain is keeping tabs on what the actual duration is. Somewhere in there, it would seem, is a clock that tracks the pace of "real" time and notifies us when our experience veers from it. But many scientists are starting to wonder if that's really so. "The brain doesn't code for physical time, only for subjective time," one prominent psychologist told me. The notion goes back at least to William James, who contended that we can't report actual duration but only our perception of duration. The revised explanation for the oddball effect seems to bolster that argument. The oddball image doesn't seem to last longer than normal, it only seems to last longer than the image that follows it; your assessment of a duration isn't made in isolation but only in comparison to the duration of another stimulus.

"There might not be a sense in which we can understand duration in its pure form," Eagleman said. Our duration clock, like any clock, is meaningful only in relation to another clock. "You can't ever know the difference between a temporal dilation and a temporal contraction. All you can ask is a relative question: which felt longer? We don't ever know which is the 'normal' one."

He had begun an fMRI experiment to explore the idea, so I volunteered for it. Functional magnetic resonance imaging is a technology that monitors the flow of oxygenated blood through a subject's brain. The research volunteer lies still while performing a mental task and the fMRI reveals roughly which regions of the brain are involved. In this experiment, subjects would be given a basic version of the oddball test. I'd be shown a series of five words, letters, or symbols, such as "1 . . . 2 . . . 3 . . . 4 . . . January," and asked whether an oddball had appeared, while the fMRI would detect whether my neurons became more or less active during the oddball itself. I'd probably experience a duration distortion while in the fMRI, Eagleman said, but I wouldn't be asked about it. My neural response was what mattered, not my conscious response.

The fMRI lab was down the hall. An attendant was watching over a computer console; just behind it was a long window that looked into the room with the machine. I was asked to empty my pockets of anything metal; I handed over my pen, some loose change, and my father-in-law's watch. The experiment would take forty-five minutes or so, during which I would lie motionless in a tight space. It occurred to me that, in addition to whatever mental energy the experiment required, I would need to try to remember as much as possible of what was about to happen, since I wouldn't be able to keep a written record of it. I mentioned to the attendant that I'd never been in an fMRI before.

"Are you claustrophobic?" she asked.

"I don't know," I said. "We'll find out."

The machine had a round opening, from which extended a long metal tray. I lay down on it; the attendant gave me a set of headphones and a remote-control clicker for my right hand and placed a semicircular cage, like a catcher's mask, over my face. She put a sheet over me for warmth and then stepped out of the room and pressed a button, and I slid headfirst into the tube.

The enclosure was barely wider than my body. All around and through me I felt a steady pulse from the magnets driving the machine. The thought struck me, and persisted, that I was in a kind of womb. Inside my catcher's mask, about three inches from my eyes, was a small mirror; it was angled so that, without moving, I could peer periscope-like past the top of my head to the far end of the tube, where I could see a computer screen, blank and white—a light at the end of a tunnel. It was disorienting, and strange notions flitted through my mind. I was standing on my head with a distant view of the ground. I was peering out through a porthole. I was someone's homunculus, peeking out through the iris. The only sound was an electronic flutter or flicker, as from an old film projector, and for a moment I wondered if I was watching a silent film or maybe a home movie of my own past.

Then something went wrong. The software program running the experiment crashed; instead of a white screen I was now looking at a blue computer desktop littered with programming code. The calming voice of a grad student came through my headphones assuring me that this would only take a moment. A cursor jumped around on the screen, typing in an incomprehensible language of letters and symbols. Suddenly I had the very real sensation, both captivating and creepy, that I was peeking into the programming substrate of my own mind. I was HAL in *2001* watching a human try to fix me. Or there was nothing at all wrong with me, and there was no grad student; I was merely reflecting on my situation and, through some fortuitous glitch, had lifted the curtain on the machinery of reflection.

The screen went white again and the experiment finally began. One by one individual words or images appeared: "Bed . . . Sofa . . . Table . . . Chair . . . Monday." Then "February . . . March . . . April . . . May . . . June," and so on. After each series, a typed question appeared on the screen: "Was there an Oddball?" My task was to press a button on my remote control—left if yes, right if no—to indicate whether or not I'd seen something that violated the broader category. The procedure repeated, again and again. The five words or images appeared in quick succession, followed by an extended pause before the question was shown, during which the screen went white. I'd been instructed not to push the yes-no button until the question appeared. As I waited, I found myself falling into the blankness. It lapped against my memory, loosened my grip on the past, such that by the time the question appeared I struggled to recall the words or images I'd seen only moments before. *Oddball*—what does *oddball* mean again?

I drift off. As soon as the series ends and the white pause arrives, I put my finger on the correct button just in case I won't remember what to do when the moment comes to choose yes or no. As each image appears, it looms large and feels ever present, then is gone from me. I would say that I am lost in the now, but a trickle of thoughts wends from or toward a vague future: I'm a little hungry; the headphones are starting to hurt my head; my feet are falling asleep; how many more questions will there be? I fall asleep. I wake up. Is this an afterlife? I may be giving birth to something, or something may be giving birth to me: an idea, a code, a word.

At last I'm withdrawn from the metal tube and I realize I am my-self again, fully clothed, in a laboratory in Houston. The lab attendant removes the sheet and flips open the cage from across my face. On my way out she hands me a CD containing a hundred-odd black-and-white images of the inside of my head: this is your brain on time. In a few months, after Eagleman has run a few dozen subjects through the fMRI and analyzed the data, the results will mean something. For now I am a data point in a gathering set of data points.

"Congratulations," the attendant says, brightly. "You're part of the family now!"

What if time is little more than another color?

Eagleman has come around to thinking that the perception of time, at least on the scale of milliseconds, is a matter of coding efficiency. Your assessment of how long a stimulus lasted is a direct function of the amount of energy your neurons spent to process it; the more energy it takes for your brain to represent something, the longer the event seems to have lasted.

Oddballs are one line of evidence. As you view a series of identical images, the amplitude of your neurons' response decreases; they ex-pend less energy reproducing the same image again and again. Those images register as having a shorter duration—but you're unaware of it until an oddball image appears, which seems to last longer by compar-ison. In search of more evidence, Eagleman gathered up every journal article he could find on the subject—about seventy different studies that involved durations of one second or less. They all seemed to bol-ster his hypothesis. Suppose a dot appears briefly on a computer screen and you're asked to judge its duration: the brighter the dot, the longer it will seem to have lasted. Likewise, a larger dot seems to last longer than a smaller one; a moving dot seems to last longer than a stationary one; a fast-moving one seems to last longer than a slow one; one that flickers quickly seems to last longer than one that flickers slowly. In general, the more intense a stimulation is, the longer its perceived duration. Likewise, higher numbers seem to have longer perceived durations than lower ones; if you're shown a numeral such as "8" or "9" for about half a second, it will seem to last longer than a lower number, such as "2" or "3," that is visually the same size and shown for an identical duration. Brain-imaging studies showed parallel results: a bigger object triggers a larger neural response in the viewer than a smaller object; something brighter triggers a bigger neural response; objects that are moving or flickering faster or looming all trigger a larger neural response. Time—duration—seems to be the brain's way of expressing how much energy it's spending on a task.

In that respect, duration may be very much like color, Eagleman said. Colors don't physically exist in the world; rather, our visual system detects certain wavelengths of electromagnetic radiation—a narrow spectrum of them, for that matter—and interprets these as red, orange, yellow, and so on. "Redness" isn't bound to a red apple but instead arises in the mind, as a translation of the energy radiating from the object. Maybe duration is likewise painted on by the mind. "In the lab, we can make something seem to last longer or shorter because there is no sense in which time is a 'true thing' that your brain just passively records," he said. The notion that time may be no more real than color "sounds completely crazy," he conceded. "Obviously, if someone were to hear that, they'd say, 'Well, what about the trajectory of my sense of self? What about the narrative of my life?'"

One afternoon I hopped into Eagleman's pickup and we headed toward Dallas, to the Zero Gravity Thrill Amusement Park, a four-hour drive away. Before long, we'd left the suburbs of Houston behind and entered the Texas flatlands: arid, brown, empty of everything but truck stops and fast-food restaurants. At one point we passed a large wooden sign that read, "Lost: the Map is my Book." Or was it "the Book is my Map"? We sped past it doing eighty.

The freefall experiment has since become something of a trademark for Eagleman. The idea is straightforward: the volunteer is placed in a situation—in this case, a controlled free fall—frightening enough to make time seem to slow down, and Eagleman tries to measure what "time slows down" actually means. It's a recapitulation of his childhood accident but it also gets at the film metaphor: when time slows down, how broad is that perception? By this point I had read and heard numerous personal accounts of time standing still. Even my mother told me a story: she was driving on the highway one day when a refrigerator fell off a truck directly in front of her and she swerved, seemingly in slow motion, around it. But nothing like that has ever happened to me. For $32.99 plus tax, the Zero Gravity free fall seemed like a safe and easy way to access this allegedly profound and psychedelic experience, so I signed on to do it.

The key to the experiment was a wristwatch-like device that Eagleman had designed; he called it a "perceptual chronometer." It had a large digital readout, but instead of the time it displayed a number and its negative image, in rapid succession, like this:

When the rate of alternation is relatively slow the wearer can make out the number, but as the rate increases past a certain threshold the images perceptually overlap and cancel each other out, so the wearer sees only a blank screen. The threshold is different for every participant; before the free fall, Eagleman determines what it is for each individual and sets the alternation rate a few milliseconds faster. I would wear the device and watch it as I fell. If indeed time slowed down, I should be able to perceive more per unit of time and correctly report the digit on the readout.

The amusement park was a few miles outside Dallas, past a strip of gas stations and along a dirt road lined with young trees coming into leaf. As we approached I could see, above the treetops, the upper half of a spindly metal structure; it vaguely resembled the Eiffel Tower but was much shorter and painted blue. Eagleman noticed me taking notes and provided a narrative voice-over: "They pull down a narrow dirt road, a tower in the distance . . ."

I'd envisaged a large, crowded amusement park, on the scale of Six Flags. But there was only a small white building, where tickets were sold, and behind it five thrill rides. The largest was the blue tower I'd seen from a distance—the Nothin' but Net 100-foot Free Fall attraction. We'd come on a Friday afternoon and only two other people were there—two young men, clearly identical twins, with bright smiles and hair cropped short. One was getting married the next day.

When Eagleman and his colleagues first began thinking about how to design the experiment, they went to Astroland and rode all the roller coasters, but none seemed scary enough to induce a duration distortion. At a glance, this one didn't look very scary to me either, although the other attractions did. The Texas Blastoff was a giant slingshot: a metal sphere, large enough to hold two seated passengers, that hung between two fifty-foot poles on thick, Bungee-like cords; the sphere would be winched to the ground and then launched into the air and allowed to bounce and spin. The Skyscraper resembled a two-bladed windmill a hundred and sixty-five feet tall; there was a single-passenger capsule at the end of each blade, and the whole thing would whirl at an unpleasant rate. Nothin' but Net looked sedate by comparison: a small, square platform at the top, two hundred feet up, and two nets, one under the other and fifty feet off the ground, stretched between the tower's four legs.

"One thing I'll mention is that it's totally safe," Eagleman said. Somehow, until then, the issue of safety had never seriously crossed my mind. "I looked at all the statistics on these things; there's never been an accident."

We sat on a picnic table and watched as the twins went. They stood on the platform, wearing harnesses, with the ride operator, a muscular man in a T-shirt. The platform had a square opening in the middle; the operator attached a cable to the front of one of the twins and carefully positioned him, his back to the ground, over the opening and then through it, so that he dangled just beneath the platform. Then the twin dropped—plummeting like a small boulder into the net, which billowed from the impact. A couple of minutes later the other twin dropped. Eagleman suggested that I guess how long their falls had taken, and he wrote down the numbers: 2.8 seconds, 2.4 seconds. When the brothers were done, they wandered our way, their eyes wide. "The fall is longer than you think," one said.

It was my turn. The nets had been lowered to the ground to let the twins off. Then the platform descended, like an elevator, and I stepped on. The operator helped me into the body harness, which was surprisingly heavy; it was carefully weighted to ensure that I didn't roll as I fell, he said, so that I would land in the net on my back, semireclined. He clipped my harness to a railing to prevent me from accidentally falling before the net was in place. Then the platform shuddered and began to rise. As we neared the top, the cables that lifted the platform began to creak in an unnerving way, and we swayed slightly in the breeze. I suddenly remembered that I am terribly afraid of heights; I looked around and up, anywhere but down through the hole in the middle of the platform. A half mile away, I could see bulldozers and other earth-movers kicking up clouds of gray dust in a gravel quarry. In the other direction, across the road, a go-kart course was under construction, and beyond that was the freeway.

The platform came to a halt. The operator unclipped me from the railing and then clipped a cable from overhead to the front of my harness. His movements were quick and clinical, like those of a hangman. He proposed that I let go of the railing, and I noticed that my hands were clenching it; it took me several moments to release my grip. He instructed me to stand with my back to the opening and then to lean backward over it and let my weight pull the cable taut. I might have been a kid on a tire swing except I was two hundred feet up, trembling in the breeze.

With great care, he positioned me over the opening and slowly lowered me through it. He made some final adjustments. I dangled and looked up—at my metal umbilical cord; at my hands, which gripped it; at the sky. Maybe because I could no longer see the ground, my terror had subsided slightly. The pull of gravity was powerful, almost magnetic, and strangely reassuring.

I would have to let go of the cable, the operator said. Against every instinct, I did. I made a fist with my left hand and held it firmly with my right hand; that way, Eagleman's device, which was strapped to my left forearm, would be squarely in sight as I fell. I fixed my gaze on its little window, where, allegedly, some number whirred at me—positive, negative, positive, negative, too fast for me to make it out—and waited to return to earth.

And then I am falling.

I remember the metallic click as my harness clip is released from the cable. But that sound comes after. First and immediately I am being pulled, almost yanked, downward, as if I were attached to an anchor or a dead weight that had been tossed overboard, and then I realize that the weight is me: I am the sinking anchor.

Only then, after the fact, does the click of the harness clip register to my ear. I am untethered. A tightness wells in my stomach, induced by my downward acceleration. It intensifies and dizzies me; I fear that it will never stop but rather grow and crush me from within. "I see that time is a kind of strain or tension," Augustine wrote. "And I would be very surprised if it is not the tension of consciousness itself." I have no thoughts; I am all tension, pure weight.

My own definition of the present is unscientific: just enough time to realize you're thinking about the present, by which time you've moved on to the next moment. As I fall, the sensations accumulate—the click of the harness release, the leadenness of my body—and I sense my consciousness piecing them together, coalescing around a word or term that will capture the situation. A thought is emerging, it almost exists, it's . . . *How long this will last?* And with the finality of cement, it ends. I hit the net, sink in deep, and am lowered to the ground.

I didn't feel well on the drive back to Houston. My neck hurt from the landing in the net, which wasn't as soft as I'd imagined, and my head ached. I was thirsty. Frankly I felt deflated. Once, years ago, I went skydiving. I vividly recall the feeling of pure terror as our very small plane puttered up to fourteen thousand feet, a motorboat in the sky; the faith it took to roll out the door into empty air; and then the leisure of the fall, which at terminal velocity felt like no fall at all. Somehow I'd envisaged a similar experience in Dallas: a slightly more than ephemeral chance to take in my surroundings, to watch the sky recede. Now it was over, with so little of it to remember.

Eagleman had instructed me to keep my gaze on the chronometer as I fell, to try to make out the number on its face. Now he asked about it.

"Hey, yeah, did you see the number?"

I didn't. Glare from the sun had washed out my view of the readout, or maybe I'd held my arm at a bad angle. Eagleman had already run the experiment on twenty-three volunteers—a very small sample, he readily admitted. On average, they reported that their own falls had lasted thirty-six percent longer than the falls they'd watched. But none of them had been able to read the number on the wrist device.

"People were not able to see in slow motion, which you'd be able to do if visual perception was like a video camera," he said. "If you were slowing down time by thirty-five percent—if you were slowing down a movie camera by thirty-five percent—you'd easily be able to read those numbers off the screen at the rate we were presenting them. You can distort duration, but it's not 'time' that slows down."

Why, then, did my own fall seem to have lasted longer than the one I'd watched? I assumed that adrenaline is involved, but adrenaline works relatively slowly, Eagleman noted: first the endocrine system gets notified, which releases hormones that trigger the adrenal glands to release their hormones. The more likely factor is the amygdala, a walnut-size region of the brain that is involved in recording memories, especially emotional ones. Neurons from the eyes and ears plug directly into the amygdala, which can then shout messages to the rest of your brain and body. The amygdala is a megaphone, amplifying and relaying incoming signals to draw immediate attention; it can respond within a tenth of a second, faster than higher brain regions such as the visual cortex. If you see a snake or an even shape that resembles one, the amygdala sounds the alarm, enabling you to jump before you realize what you've seen. And because the amygdala is connected to all parts of the brain, it also can act as a secondary memory system, laying down memories in particularly rich form.

A body in free fall "is in total panic mode, it goes against every Darwinian instinct you have," Eagleman said. "Your amygdala is screaming." Your sensations of the event, though fleeting, pass through the amygdala, where they gain added texture as they're pressed into memory; it's a bit like recording video in high resolution instead of standard grade. In retrospect, as you stand on the ground reflecting on the fall, the added richness of the memory creates the impression that the fall lasted longer than it did. Whether that duration distortion is helpful—whether in fact you can respond more quickly or wisely—is much harder to say. "There are a lot of things that the experiment cannot rule in or out," Eagleman said. "But what it can rule out, at least, is that the whole world slows down like a movie camera. We have no evidence at this point that that can happen."

In the Beginning

Adam is ten months old, a solid baby with large brown eyes. He sits in a comfortable high chair in a small, darkened, nearly soundproof room in a psychology laboratory and is gazing back and forth between two computer monitors arranged side by side on a tabletop in front of him. A video is playing on each screen. Both feature a woman's face that looks directly at Adam and speaks slowly. It's the same woman in both videos, with the same smiling, bright-eyed expression, but there is no audio, only the movement of her lips.

Occasionally, for reassurance, Adam looks over at his mother, who sits quietly nearby. Between the two monitors is a small camera aimed at Adam, which streams a live video of his face to a desktop monitor just outside the room, where two lab assistants and I watch Adam's eyes dart around and his expression shift from engaged to wary to bored to curious again in the span of a few seconds. Behind our monitor is a one-way window with a direct view of Adam in his chair. The setup has a funhouse quality to it: we are watching Adam through the window as he watches the two faces on his monitors, and simultaneously watching his looming face on our screen. From time to time Adam looks directly at the camera and I have the brief, eerie feeling that he is watching us or knows that we are watching him. His gaze soon shifts back to the two faces in front of him; he peers, he points, his eyebrows lift. In the low light, in his five-point harness, Adam looks a bit like a pilot or an astronaut peering forward into space.

The lab belongs to David Lewkowicz, a developmental psychologist at Northeastern University. For the past thirty years, Lewkowicz has sought to understand how the budding mind orders and makes sense of the sensory information that floods in from the moment of birth and even earlier. How does the brain track the arrival times of different bits of data, and how does it integrate them to provide us with a unified experience? How does it know which properties and events belong together in time? The power and subtlety of this ability is evident in the two videos in front of Adam. To an adult viewer, it's immediately obvious that the woman is saying something different in each video, even when there's no audio; the lip movements on the two screens aren't the same. After a few moments the sound comes on and the woman's voice becomes audible. "Get up," she's saying in a singsongy way. "Get up right away. Today we're going to have oatmeal for breakfast! Then we'll have time to putter around the house . . ." The monologue matches the talking face on the left; I pair the audio and video so intuitively—the sound and lip movements fall into instant synchrony—that my attention is immediately pulled in to her patter; the other face might as well not be there. Sometimes, instead, a different monologue becomes audible—"Are you going to help me fix up the house today?"—and I immediately link it to the screen on the right. Some trials use the bright, smiling face of a different woman, and her voice comes up in Spanish. So powerful is an adult's ability to detect synchrony that, even without comprehending the language, I know which of the two sets of moving lips her words belong to.

Does a baby have the same ability? It would seem unlikely. A newborn doesn't hear well, can't visually focus on anything much more than a foot away, and has limited experience in the world. "The baby, assailed by eyes, ears, nose, skin, and entrails at once, feels it all as one great blooming, buzzing confusion," William James proposed in 1890. Maybe. But Lewkowicz has found that infants begin to draw order from the swirl surprisingly early. He has run the talking-face experiment on hundreds of babies and toddlers. They watch the two faces side by side, the lips moving in silence, for a minute. Then the audio comes on, and the researchers watch the screen to see whether the baby's eyes linger on one talking face more than the other. With remarkable consistency, children as young as four months express a preference for the face that actually matches the voice—despite the fact that the child has never seen the face before, doesn't understand the words, and may not even be familiar with the cadence of the language.

Instead, Lewkowicz contends, he makes the correct match by the simplest means, by matching the starts and stops of the audio stream to the starts and stops of the visual stream. The infant grasps synchronization, recognizing when things happen together in time. *Soon* will come soon; *eventually*, later. But first, and from a very early age, humans know now from not-now, and that distinction is sufficiently powerful to kick-start our sensory development. "You're legally blind," Lewkowicz says. "Your hearing hardly works. Either it's a blooming, buzzing confusion, or you've got some basic, really primitive mechanism that gets you up and running. And that mechanism is synchrony."

In 1928 Europe's leading physicists, philosophers, and natural scientists gathered in Davos, in the Swiss Alps, for a conference and an exchange of ideas. Until then, the alpine resort had been known mainly as a sanatorium, a clear-aired refuge for recuperating minds and failing bodies. Hans Castorp, the protagonist of Thomas Mann's 1924 novel, *The Magic Mountain*, goes to Davos to visit a tubercular cousin; he slips into the languorous pace of the mountains, glances at his pocket watch, and considers the subjective nature of time as Mann has distilled it from Heidegger, Einstein, and other contemporary thinkers. Castorp wonders why miners, trapped in a cave, emerge after ten days thinking that only three have passed. Why do "interest and novelty dispel or shorten the content of time, while monotony and emptiness hinder its passage"? What do we make of a man who allows himself the habit of saying "yesterday" for "a year ago"? Also, he asks, "Are hermetically sealed preserves on the shelf outside of time?"

By 1928, with tuberculosis and the hospice business in decline, Davos began to repurpose itself as an intellectual spa. Einstein was invited to preside over the first Davos conference. Gandhi gave a talk, as did Freud. Also speaking was the Swiss psychologist Jean Piaget, who at thirty-one was already known for his studies of how children come to understand the world. Piaget had taken an active interest in the natural world as a child; his first scientific observation, at age eleven, was a sighting of an albino sparrow—or, as he cautiously phrased it, "a sparrow presenting all the visible signs of an albino." He began a career as a zoologist, studying mollusks, but soon was preoccupied by the question of how a child's thinking develops over time. He proposed that we are born into the world with our five senses disconnected from one another; only through experience—by touching, biting, playing, and otherwise interacting with things—do these senses begin to overlap and intercommunicate. Gradually we learn which input goes with another, and we develop a rich understanding of what any particular object "is": a spoon has such-and-such an appearance; it feels like so to the touch; it makes a certain sound when banged on the table. Many of the examples Piaget offered were derived from detailed studies of his own children. He ran simple experiments with them, took careful notes, and knew almost on a daily basis which perceptions were coming online. Nowadays his key insights are taken for granted: that children perceive the world differently than adults do, and that their perception coheres as their senses mature and integrate, a process that takes years.

After Piaget gave his talk, Einstein approached him with a series of questions. How, the physicist wondered, do children come to understand duration and speed? Velocity is typically defined as a function of distance over time—meters per second, or miles per hour. Is that how a child first conceives it? Or is its notion of speed more primitive and intuitive? Does a child grasp speed and time together, or one before the other? Does the child understand time "as a relationship, or as a simple and direct intuition"? Piaget went off to investigate, and his studies became the basis for his 1969 book *A Child's Conception of Time*. In one experiment, involving children four to six years old, he placed before his subject two tunnels, one of which was obviously much longer than the other. Then, with a metal rod, Piaget pushed a doll through each tunnel in such a way that both dolls arrived at the far end of their respective tunnels simultaneously. As Piaget described it, "We ask the child: Is one tunnel longer than the other?"

"Yes, that one."

"Did both dolls go through the tunnels at the same speed, or did one go faster than the other?"

"The same speed."

"Why?"

"Because they arrived at the same time."

Piaget ran many iterations of the experiment. He used windup snails and toy trains and even ran around the room with the child. Scientist and subject would start at the same time and stop at the same time, but Piaget would run a little faster and leave the child a few feet behind. "Did we start together? *Yes.* Did we stop together? *Oh, no.* Which one stopped first? *I did.* Did one stop before the other? *I did.* When you stopped, was I still running? *No.* And when I stopped, were you still running? *No.* So did we stop at the same time? *No.* Did we run for the same length of time? *No.* Who went on longer? *You did.*" This exchange was typical, Piaget found. A young child might grasp simultaneity—two people starting and stopping at the same time—but if Piaget and his subject traveled different distances, the child conflated physical length with duration. Time and space, speed and distance, were all of a piece.

Piaget's work made clear that what we adults sometimes refer to as our "sense of time" in fact has many facets, which don't all emerge at once. "Time, like space, is constructed little by little and involves the elaboration of a system of relations," he concluded. In the decades since, developmental psychologists have unraveled time into several strands, including one's grasp of duration, of rhythm, of order, of tense, and of time's unidirectionality. In one experiment, William Friedman, a psychologist at Oberlin College who has written nearly as much as Piaget did about children's perception of time, showed eight-month-old babies a video clip of a cookie falling to the ground and breaking apart. The infants found the video far more compelling when Friedman ran it in reverse, which suggests they had some sense of time's arrow: they knew a strange sight when they saw it.

By the time a child is three or four, a sense of chronology has begun to settle in. Katherine Nelson, a psychologist at the City University of New York, found that her young subjects could answer vague questions—"What happens when . . . ?"—with surprising accuracy. Most understand that making cookies involves putting dough in the oven, taking them out, and eating the result, in that order. Show a young child a picture of an apple and then a picture of a knife and she will correctly pick a picture of a sliced apple as the next image in the sequence.

By age four or so, a child has a relative grasp of how long common events last: watching a cartoon takes longer than drinking a glass of milk, and sleeping at night takes longer still. If she hears a sound for as long as fifteen seconds, she can accurately reproduce its duration. But the past and future are more confounding. A child typically speaks in the correct tense by age three but may not grasp the difference between "before" and "after" until age four. Ask a four-year-old at what time of day class met seven weeks ago and most will say "morning" but won't recall the correct season. Ask a five-year-old boy in January which will come sooner, Christmas or his birthday in July, and he'll likely say Christmas. At this age, Friedman found, past events reside in the mind as if on islands of time—distinctly there but not yet related to one another or part of a larger archipelago. The landscape of future events is even more inchoate, though not unpredictable. By age five, Friedman found, children understand that animals grow, not shrink, and that a gust of wind will send a neat pile of plastic spoons whirling into the air but won't stack them up again.

These degrees of temporal knowledge are mostly learned, psychologists say—absorbed as we grow into our social lives. If a six-year-old is shown a group of cards depicting various events from a typical school day, she can place them in correct chronological order or even put them in reverse order. By seven, she can correctly perform a similar task involving seasons or holidays over the span of a year, but only in forward order. Organizing time in reverse—"If it's August and you go backward in time, which will you come to first, Valentine's Day or Easter?"—doesn't come easily until she's at least a teenager. The disparity reflects her accumulated experience, Friedman believes. By age five, a child has repeated the events of a typical day—waking, breakfast, lunch, snack, dinner, story, bed—hundreds of times, whereas her encounters with months and holidays (that is, days sufficiently distinct from one another that they merit individual names) are still relatively few. It takes time to grasp time.

And how we learn about time affects our facility with it. One reason that young children have a hard time thinking about months and weekdays in reverse order, Friedman found, is that early learning is often list-based. We learn the weekdays and months as a series—"Monday, Tuesday, Wednesday, Thursday"—much as we do with the letters of the alphabet. Answering a question like "Which comes first, February or August?" is a matter of simply running down the list internally. (Studies have shown that young children often move their lips while solving such problems.) But we only learn these lists in forward order; it can take years, until our teens, to loosen the items from the schema sufficiently to learn their relationships in reverse. Culture and language play a part too. A study of American and Chinese second- and fourth-graders found that the Chinese children had a much easier time answering questions such as, "Which month comes three months before November?" That's because, in Mandarin, the weekdays and months are named numerically; November is "month eleven." A question about temporal order requires American children to manipulate words on a memorized list; for Chinese students, it's a math problem and is quickly solved.

Lewkowicz discovered Piaget as a senior in high school. In 1964, when he was thirteen, his family emigrated from Poland to Italy and from there to the United States. When they landed in Baltimore, he spoke no English. He remembers his first few years in the United States as socially bewildering but much less adversarial than in his neighborhood back home, where Jews were unwelcome. By his senior year in high school he was working as a lifeguard; he found it boring but liked taking in the scene from some remove. He read Piaget, which opened his eyes to the study of the psychology and behavior of children—to the understanding, Lewkowicz says, of "where it all comes from."

Lewkowicz is slender and hale, his hair beginning to turn silver; only occasionally does his voice betray an Eastern European accent. More than once during my visit to his lab he exclaimed, sincerely, "I love what

I do!" At one point we joined two grad students who were reviewing the video from an experiment they had run earlier in the day. The face of an eight-month-old baby loomed on the monitor, its eyes wide. "Our data are right there," Lewkowicz said enthusiastically. "The eyes are our window to everything. All we do is measure where they look." Babies may not speak, but they do express a quantifiable metric with their gaze. In Lewkowicz's experiments, which follow a common protocol, he shows the baby something on a computer monitor again and again until the child loses interest and looks away. The researcher, watching the baby's eyes remotely, clicks and holds a computer mouse when the baby looks at the screen, then unclicks when his gaze shifts; the length of the click provides a measure of attention. When the child's attention span drops below a certain threshold after three trials, the computer automatically shows the child a new stimulus on the screen.

"The baby is in charge," Lewkowicz said. "It gets to tell us what it wants to see. And it gives us a hint of what's going on in the brain. Babies tend to search for novel things; they're constantly seeking new information, foraging for novelty. What we do is bore the heck out of them. We show them the same event over and over, and then we change some aspect of it to see if they detect the change; if they do, that suggests they learned the original event. We just keep our finger on the button to measure how long the baby is looking. It's easy to do and very powerful."

Among researchers who study the perception of time, infants are still something of a frontier; Friedman has described early childhood as "a kind of wasteland for students of cognitive development." But the advent of computers and eye-tracking equipment has made it easier to probe those early weeks and months of life and to begin to understand what humans know about time when we enter this bright world. For instance, it's been shown that infants as young as one month can distinguish between phonemes like "pat" and "bat," which can differ in duration by as little as two-hundredths of a second. Another study found that two-month-olds are sensitive to the order of words in a sentence; if an audio track of a sentence such as "Cats would jump benches" is repeatedly played for the infant and the sentence is suddenly changed to

"Cats jump would benches," the baby's attention perks up. At one point Lewkowicz walked me through an experiment that involved different shapes—triangle, circle, square—falling one by one from the top of a computer screen to the bottom, each landing with a distinctive bonk or beep or bing. He would show his subjects, all four to eight months old, a particular sequence until they grew accustomed to it and lost interest; then he would show them a new sequence—the same shapes and sounds but in a different order—to see if they would notice. Almost invariably they did, which to Lewkowicz revealed a fairly sensitive awareness to temporal order.

"There really isn't a huge literature on this," Lewkowicz said. "There's been this explosion in research in early cognitive development on all sorts of topics, but time has not been one of them. Yet it's such a basic feature of this world." He added, "Babies live in a very different temporal world than we do, I think. I'd love to get into their heads and come back out."

As an undergraduate, Lewkowicz was part of a team studying the sexual behavior of octopuses and helped build the world's first laboratory aquarium that was capable of keeping the animals alive indoors. As a grad student, he worked in a neonatal intensive-care unit, studying whether the unit's twenty-four-hour lighting and constant noise might affect the newborns' development. He wondered, among other things, why ninety percent of infants in a nursery lie with their heads to the right. (It's still an open question; some researchers think it may relate, or even contribute, to the prevalence of right-handedness in our species.) And taking a page from Piaget, he began to look at how the human mind, even very early in its development, begins to integrate the sensory information flooding its system.

For the first two to three months of life, human infants are subcortical animals. The cerebral cortex—the several rich outer layers of neurons that help the brain organize perceptions and provide a foundation for abstract thought and language—hasn't yet come online or begun to influence or inhibit the many basic functions of the nervous system. Once the cortex kicks in, an infant begins to smile; it is as if he has finally wakened to the world. Until then, Lewkowicz said, "you get the feeling they're not plugged in." One of his early experiments showed that, in those initial weeks, an infant organizes its sensory world based not on the type of information flooding in but simply on its quantity. Grown-ups have this ability; if you show an adult small patches of light of varying brightness and then have her listen to sounds of varying loudness, she can match them according to their amplitudes: this light is about as bright as that sound is loud. Lewkowicz found that three-week-old infants can make similar connections.

"At birth, babies can link auditory and visual information at a very rudimentary level, in terms of intensity, the amount of energy," Lewkowicz said. "What it suggests is that babies have a basis for building their world, if you will; they use simple mechanisms for getting going and figuring out what goes with what."

At some point in his research Lewkowicz began thinking about faces. An infant can't see clearly beyond a foot or so, but the one foreign object it encounters regularly is the face of its caregiver. This face is a complex stimulus, with moving lips and shifting expressions and producing an ever-changing patter of sounds. Lewkowicz revisited Piaget's question about intersensory integration: can a baby perceive a talking face as a coherent object? When and how does that perception begin to emerge, and what attributes make it possible? Lewkowicz soon realized that much of what a talking face conveys to an infant—an organism with no grasp of vocabulary or linguistic content—involves time and timing. When the face opens its mouth, it makes a sound of some duration; it speaks quickly or slowly; it speaks or sings with a rhythm, which for infants is a powerful tool for organizing information into meaning. A baby may learn the rhythm for "Twinkle Twinkle Little Star" long before he understands that "Twinkle" is a separate word, much less what it means. (As preschoolers, my boys were eager to know how to write the letter "LMNOP.") Eventually, the infant perceives that the motion of the lips synchronizes with the sounds. One spoken sentence holds many dimensions of time, and a newborn primed to learn about them has a ready instructor staring it in the face.

One morning in his office, Lewkowicz was complaining about his local cable provider. He had tried to watch a documentary the previous evening but the audio and video streams were so badly out of sync that he finally turned off the television in frustration. "The audio was over by the time the person started talking," he said, exasperated. It happens occasionally on every channel carried by his cable service and to every local subscriber that Lewkowicz has met. The problem also describes his research interest in a nutshell.

The perception of time has many facets, but perhaps the most critical is synchrony—our grasp of whether separate sensory streams, such as the sound of a voice and the sight of someone's moving lips, are occurring simultaneously or not, and whether they belong to the same event. We're highly attuned to it; studies have found that as you watch a film clip of someone talking to you, you'll notice if the audio and video streams go out of sync by as little as 80 milliseconds, or less than a tenth of a second. And if the audio track lags the video by 400 milliseconds— not quite half a second—you'll have a much harder time understanding what's being said.

Perceptual unity serves us well, and the mind works hard to achieve it, often at the expense of strict accuracy. Studies in the nineteen-seventies revealed that if you see a visual stimulus—the moving mouth of a puppet, say—in one part of the room and simultaneously hear a sound that originates elsewhere, the sound will seem to occur closer to the visual input than it actually does. That's the ventriloquism effect, and it's the slippery power of intersensory integration at work. A voice isn't even required; a few plain tones and a sock puppet, separated by some distance, will induce the perception of unity.

A related illusion is the McGurk effect: the tendency to mix up audible and visible syllables if they are perceived simultaneously. For instance, if you watch a person on a video say "ga" while the audio track dubs in the syllable "ba," you'll almost certainly hear something that is neither: the syllable "da." The McGurk effect also can be induced by touch. In a Canadian study, subjects listened to a voice say four sounds:

the aspirated syllables "pa" and "ta," which are produced with an inaudible puff of air, and "ba" and "da," which are unaspirated. When the scientists simultaneously delivered a small puff of air to a subject's hand or neck, the listener heard "pa" instead of "ba" and "ta" instead of "da"—as if the puffs had been heard, not felt. The effect is so consistent that researchers now wonder whether hearing aids could be outfitted with airflow sensors, so that a hard-of-hearing listener might, in effect, hear with her skin.

The brain works hard to stitch the incoming data into a coherent representation of the world. As an adult, you can recognize that the voice and the pair of lips on your television are out of sync because you have extensive experience with voices and moving lips; you know that they tend to work in concert, and you understand the words and ideas that emanate from them. But babies have none of that experience, Lewkowicz noted, and no assumptions. To watch an infant watch a talking face is to access a very different understanding of the present. Lewkowicz built a study protocol around it, which he calls the talking face experiment.

In his office, he showed me a short video of a woman's face on his computer monitor. At the start of the video clip the woman's mouth is closed; then slowly and clearly she says the syllable "ba" and closes her mouth. Lewkowicz showed the video to a series of babies ranging in age from four to ten months old. Each one watched the video over and over until its attention wandered off, at which point the video changed. Then the same woman said the same syllable, but this time the audio and video were out of sync; first the "ba" sound came on, then after 366 milliseconds—a third of a second—her lips began to move. To an adult the asynchrony is egregious, but the infants didn't register it. Nor did they notice anything amiss when the two streams were as much as a half second out of sync.

"They don't get it," Lewkowicz said. "They only get this one." He showed me the video again, but this time the audio was fully two-thirds of a second—666 milliseconds—out of sync. "The sound is finished before the mouth even opens!"

This interval—the brief span of time within which separate streams of sensory data are labeled as belonging to a single event—is known as the intersensory temporal contiguity window. In many respects it's a good working definition of "now," although the size of this window varies depending on the stimuli and who is observing it. For babies, "now" lasts two-thirds of a second when they're watching a talking face. But, Lewkowicz found, if they watch a more punctate event, such as a ball bouncing on a screen, they'll notice if the sound is only a third of a second out of sync; their "now" is smaller, though still demonstrably longer than an adult's, at least when it comes to integrating more than one sensory stream.

"The baby's world is a slower place, I think," Lewkowicz said. He isn't sure why. It may be that the neurons in the young brain transmit signals more slowly. The early neural system lacks myelin, a fatty material that coats and insulates neurons and speeds conduction; myelin is deposited gradually during the course of childhood, and the process can take twenty years. "The baby's brain is a slower organ, there's no doubt about it," Lewkowicz said. "But from a perceptual standpoint it's hard to think about. What does it mean to say that the baby's world is slower? From the baby's perspective it's just the world. The question is, what consequences does it have for the baby's perception of its world?"

It's a marvel that babies perceive synchrony at all. As adults, we recognize that a mouth and its voice are out of sync because we know something about words and lips and the sounds associated with them. A baby knows nothing. Indeed, when it watches a talking face, it barely looks at the mouth, at least during its first six months; instead, Lewkowicz has found, it directs almost all its attention to the eyes. Only once a baby reaches eight months old or so does it begin to consistently track the movement of lips.

How, then, does a baby know whether two sensations are in sync or not? Lewkowicz thought back to his doctoral research, which showed that newborns can effectively match two stimuli from different sensory modalities—a sight and a sound—based on intensity. Lewkowicz suspected that babies might register synchrony in a similar way. He designed a variation of the talking face experiment and, with colleagues at the University of Padova, in Italy, conducted it on babies as young as four months. The babies saw two silent videos side by side. One showed the face of a monkey silently making an O with its mouth, as if cooing: in the other, the same monkey jutted out its jaw in an inaudible grunt. When a recording of one of the two calls was played aloud, the babies consistently paid more attention to the video that matched it—that is, to the monkey whose lip movements started and stopped exactly simultaneously with the audio track. The researchers then conducted an even more basic version of the experiment; this time the babies heard not a monkey call but a simple tone that matched the duration of one of the two monkey's lip movements. Again the infants, some no more than a day old, were drawn to the video that matched the audio in duration.

To Lewkowicz, this clearly showed that a newborn's perception of synchrony has nothing to do with the content of what's being synced. What looks like infant superintelligence—the ability to pair monkey faces to voices—is little more than a mechanical circuit. The baby matches the start and stop of the audio with that of the video; its neural system is simply registering the onset and stopping of two streams of energy, like noting that a light and a noise have both switched on and off at the same time. If the two activities coincide, they define the same event. It's a bit like doing a jigsaw puzzle by just doing the edge pieces. Infants use synchrony to define the borders of events while ignoring the interior pieces—the higher-level information that might interest an adult, such as words and phonemes or a basic understanding of what lips do, which an infant's nervous and sensory systems are too immature to process.

"It's as if they don't care what's inside the stimulus," Lewkowicz said. "Just give them things that come on and off at the same time and it coheres."

Back in the soundproof room, ten-month-old Adam was providing a similar insight. On the two monitors in front of him, he watched two sets of lips silently mouth different monologues; when the audio track for one of them was played, he glanced at the lips that synchronized with it, with uncanny consistency. His choice was accurate even when the voice and the silently mouthed monologue were in Spanish, a language not spoken in his home. With a basic algorithm for synchrony—things that start and stop together belong together—he could match a voice to a face without having the least idea what the voice was saying.

In synchrony, Lewkowicz believes he has identified a core mechanism by which an infant begins to organize its sensory world. At birth, the infant's nervous system is immature and inexperienced; it can't extract higher-level information. But what it can do is detect when different sensory modalities turn on and off. We enter the world knowing nothing about monkeys but a great deal about what is happening—and what stops happening—exactly right now. "If you start with that," Lewkowicz said, "you're already starting life with a very powerful tool: things go together unless proven otherwise. It's a good way to bootstrap a coherent, multisensory world." He laughed, and added, "They do a lousy job. But it's better than James's 'blooming, buzzing confusion.'"

One might think that an infant becomes even more attuned to synchrony as it grows, but that isn't quite the case. Lewkowicz found that, by eight to ten months of age the babies in his lab can no longer distinguish a cooing monkey face from a grunting one; their efforts to match a monkey voice to a monkey face become no better than random guesses. Yet they can still accurately match a human voice to the corresponding set of human lips. As our sensory system develops, it seems to transform from a funnel into a filter, growing more selective in its choice of what to process, a phenomenon called perceptual narrowing.

"Early in development, babies are much more broadly tuned to the world," Lewkowicz said. "You have this simple device that says, 'If things go together in time, I'll put them together.' You can start to link auditory and tactile and visual information, but because it's based on energy and nothing else, you're going to make mistakes. You're going to link monkey faces and monkey vocalizations, because all you can detect is a mouth that gets bigger and smaller, and vocal onset and offset, so you're going to link them—never mind that it's the wrong species." Before long, the infant gains a working knowledge of particular faces and voices and, critically, of which faces to attend to or ignore. Experience plays a bigger role. It's the rare infant that encounters monkey faces on a daily basis, so as the neurons grow attuned to the input that actually matters, the ability to grasp the subtleties of those faces stops developing.

For similar reasons, as an infant grows its sensitivity to foreign languages also diminishes. Lewkowicz had infants from English- and Spanish-speaking homes watch two adjacent monitors: on one, a woman's lips slowly and silently mouthed the syllable "ba"; on the other, she silently mouthed the syllable "va." Then both faces were replaced by a spinning ball, and one of the two syllables was played aloud several times slowly. When the sound stopped, both faces reappeared and the researchers measured where the baby directed its attention. Regardless of their native language, six-month-old babies consistently focused on the lips that correctly matched the spoken syllable. But by eleven months, babies from Spanish-speaking homes stopped being accurate; they did no better than guess. That's because, in Spanish, the syllables "va" and "ba" sound identical; the word *vaca*, or cow, is pronounced *baca*. An infant raised in Spanish stops registering the distinction between the two syllables, whereas bilingual babies continue to distinguish between the two.

As we grow more fluent in our native environment we become less sensitive to foreign ones. Studies have shown that a very young Caucasian baby can discriminate equally well among Caucasian faces and Asian faces, but by its first year it becomes less able to recognize individual non-Caucasian faces. An infant raised in Bulgaria, where musical meters are more complex than in Western music, can discern those rhythmic subtleties as she grows into adulthood, but if she first hears them after she's a year old, she'll remain all but deaf to them forever.

Complex software programs are typically built atop simpler ones, called kernels, that do much of the basic algorithmic lifting. The ability to perceive audiovisual synchrony is something like a kernel, enabling a newborn neural network to begin organizing the swirl of sensory data without regard to its content. No prior knowledge or experience is required, only the capacity to gauge relative amounts of stimulation. With that foundation, the infant can begin to process meaning—to cope with conflicting information and to discern which sensory inputs take priority.

Lewkowicz is reluctant to call this ability innate. A prominent school of developmental psychology holds that humans are born understanding core concepts like causality, gravity, and spatial relations; these capacities have come to us through natural selection and, presumably, are based somewhere in our genes. But Lewkowicz and many of his colleagues find this argument vague and simplistic. To invoke genetics is to end the conversation there, when so many more interesting research questions might be asked. "It's a magic box," Lewkowicz said. "It's vitalism all over again."

He prefers to think of a human as an organism perpetually in development. We are beings in time. An infant is born with many basic behaviors—the ability to suckle, for instance—that soon give way to other, more advanced ones. These are ontogenetic adaptations that serve an initial purpose and then fade away. The infant's radar for synchrony may belong to this category; it jump-starts the newborn sensory system but is soon supplanted by a higher order of processing derived from experience in the waking world.

By the same measure, there's nothing physiologically magical about birth. A newborn is just the most recent incarnation of an organism that existed days and weeks before, slightly less developed, in the darkness of the womb. Studies have shown that an hours-old infant clearly prefers the sound of its mother's voice to a stranger's; one could conclude that this preference is genetically hardwired, or innate, and think up an evolutionary reason for it. (For instance, maybe natural selection favors the infant that can immediately recognize its mother.) But in fact this linguistic bond is forged in the womb, and it is acquired through experience. Several researchers have demonstrated that human hearing becomes functional in the last trimester of gestation; a fetus learns a great deal about the outside realm from the sounds that filter in. One classic study found that a fetus's heart rate quickens when it hears an au-diotape of its mother reading a poem and slows when it hears a female stranger read the same poem. A French newborn can clearly distinguish between the same story read in French, Dutch, or German, without un-derstanding any of the words. Another study found that, at two days old, the cries of French and German babies followed distinct melodies that mirrored the native languages of their mothers; they were mimick-ing sounds they had heard in utero.

Humans aren't unique in this regard. Sheep, rats, certain birds, and other animals are capable of hearing in the womb or egg. A mother Australian fairy-wren begins calling to her eggs a few days before the young hatch out. She is teaching her unborn chicks a unique begging call, which varies from nest to nest; after hatching, the chicks that can mimic it are more likely to be fed. The call is a password, enabling a fairy-wren mother to distinguish her own chicks from those of parasitic cuckoos that invade their nests.

For Lewkowicz, what seems innate at birth is just another mystery to be solved. "When you see the emergence of any kind of cognitive or perceptual skill, for me the question isn't whether it's present or not, it's 'How does it get there? When does it emerge?' If you ask me whether babies can sense time—yes, they can, but it depends on how you define time. Are they sensitive to time-based and structured information? Yes. The question is, when does this really start?"

If it seems peculiar to build a line of research around a talking face, con-sider again that, for the first few months of life, an infant's perceptual world consists almost entirely of talking faces. During the last trimes-ter, the sensory world of the fetus is limited to touch and sound. With birth comes light and motion, new dimensions to integrate. Much of what's in this new world are a parent's spoken words. The words them-selves mean nothing but, spoken aloud, they offer clues to how sights and sounds fit together; in hearing language, the newborn masters syn-chrony and learns to move beyond it. Numerous studies have shown that babies respond more strongly to a visual stimulus if it's accompanied by an audible one and vice versa; redundancy breeds salience and salience breeds comprehension.

Imagine you're at a loud cocktail party, Lewkowicz said. Someone says something that you don't quite catch, but if you watch their lips you're more likely to understand what was said. To an infant, a talking face is all redundancy. We speak slowly, lilting, segmenting what we mean to emphasize: "Here's . . . your . . . bot . . . tle . . ." The lips match the voice; even the Adam's apple jogs up and down in time. "By using rhythm and prosody and all these cues, we're allowing the infant to learn that all this stuff goes together, and to learn the word," Lewkowicz concluded. "Bingo: you've got a perfectly designed system to teach infants how to speak."

More than that, we have a system primed to teach infants about an essential aspect of time. The perception of time is many things—the perception of order, tense, duration, newness, synchrony. But time overall is one thing: a conversation among clocks, be they wristwatches, cells, proteins, or people. So how else should an infant learn about synchrony except by seeing it spoken? For new humans, at least, time begins with a word.

WHY TIME FLIES

Either the well was very deep, or she fell very slowly, for she had plenty of time as she went down to look about her and to wonder what was going to happen next.

—Lewis Carroll, *Alice in Wonderland*

This year, like every year, is flying by. It's still only July, or April, or maybe not even February, but already the mind races ahead to consider September, when school or work starts up again in earnest, as if the intervening weeks of summer had already occurred; or perhaps to June, with spring having gone by in a flash. From there it's a short mental hop to next January, whence, with some quick math, you can count all the previous Januarys in which you reflected on the year that just raced by—five, ten, so many that you've lost their details and now lump them into some broad category: "my twenties," "the years when we lived in New York," "back before our kids were born." Then it seems that your youth has flown too—or if it hasn't flown just yet, you can easily imagine a future point in time when you'll feel that it fled long ago.

How time flies: we all remark on it and have done so for centuries. "*Fugit irreparabile tempus*," the Roman poet Virgil wrote: time flees irretrievably. "Time flies, and for no man will it abide," Chaucer noted in the late fourteenth century, in *The Canterbury Tales*. From various American commentators in the eighteenth and nineteenth centuries, one hears that "time swiftly flies on anxious wings," "time rolls on with rapid wings," "time flies alas for it hath eagle wings," and "time flies, eternity beckons." Time and tide have waited for no man since before English was born. Shortly after Susan and I were married, my father-in-law took to saying, with a snap of his fingers and a bittersweet tone, "The first twenty years, they go like that!" A dozen years later, I think I know what he means. One day Joshua exclaimed, with a grand sigh, "Remember the good old days?" and he wasn't yet five. (For him the good old days involved a chocolate cupcake he remembered having eaten a few months earlier.) I surprise myself lately with how often I'm struck by this fleetingness. It seems as though there was a time not long ago when I rarely remarked, "How time flies!" Yet when I reflect back on that period of my life and compare it to the current one, I realize with shock that actual years have passed, and then I say it again. Where did the time go?

Of course, it isn't only years that fly. Days, hours, minutes, and seconds all fly by too, but not necessarily on the same wings. The brain processes a passage of time that lasts from minutes to hours differently than it treats an interval that lasts from a few seconds to maybe a minute or two. When you think back to estimate how long your trip to the supermarket took or you ask yourself whether the hour-long TV show you just watched went by more slowly or more quickly than usual, you invoke a different mental process than you do when the stoplight seems to be taking too long or when a researcher asks you to look at an image on a computer screen and estimate how many seconds it stayed there. Years are a whole other matter, which I'll get to in a moment.

Exactly why time flies "depends on what kind of time you're talking about," John Wearden, a psychologist at Keele University, in Staffordshire, England, told me. Wearden has spent the past thirty years trying to define and unravel the human relationship to time; in 2016 he published The Psychology of Time Perception, an accessible overview and history of the field. I caught up with him by phone one evening in his home as he was about to watch a championship soccer game. I apologized for interrupting. "Not a problem," he replied. "My time isn't that precious, to be honest. I'd like to pretend I was terribly busy but I'm just waiting for the football to start."

Wearden reminded me that we don't perceive time directly, as we do with light or sound. Light we perceive by means of special cells in the retina which, when struck by photons, trigger neural signals that quickly reach the brain. Sound waves are detected with tiny hairs in the ear; their vibrations translate into electrical signals that the brain grasps as audio. But we don't have special receptors for time. "The problem of the organ for time has haunted psychology for many years," Wearden said.

Time comes to us indirectly, typically by way of what it contains. In 1973 the psychologist J. J. Gibson wrote that "events are perceivable but time is not," a statement that has become foundational for many temporal researchers. What he meant, roughly, was that time is not a thing but a passage through things—not a noun but a verb. I can describe my trip to Disneyworld—there's Mickey, there's Space Mountain, there are the clouds far below my airplane window—and I can be conscious of the trip even as I take it. But I can't experience or relate a "trip" devoid of sights, activities, or thoughts. What is "reading" without words and your progress through them? Time is merely our word for the movement of events and sensations through us.

Gibson's formulation isn't far removed from Augustine's. "Do not interrupt me by clamoring that time has objective existence," Augustine wrote. "What I measure is the impression which passing phenomena leave in you, which abides after they have passed by: that is what I measure as a present reality, not the things that passed by so that the impression could be formed. The impression itself is what I measure when I measure intervals of time." We don't experience "time," only time passing.

To acknowledge and mark the passage of time is to acknowledge change—in your surroundings, your situation, or even, as William James noted, the interior landscape of your thoughts. *Things aren't as they were before.* Into the sense of *now* seeps an awareness of *then.* And performing that comparison requires memory. Time can only fly—or crawl, or leap—if you recall its previous speed: "That movie felt much longer than others that I've watched," or, "The dinner party flew by; I remember noticing the clock two hours ago but I haven't noticed it since." To the extent that time is a thing, it's a trail of your memories of other things.

"Everybody has had the experience of being engrossed in a book," Wearden said, "and then looking at a clock on the wall and saying, 'Is it ten o'clock already?' I used to think that one could measure the sense of time during the interval. But of course you can't, because you haven't felt it; it's a pure inference. That's what makes everything complicated. We talk about the feeling of time passing, but often these temporal judgments are based on an inference, not on direct experience."

Indeed, very often when we remark, "How did time fly by so quickly?" what's actually meant is some version of "I don't remember where the time went" or "I lost track of the time." The experience strikes me most often when I've driven a long distance on a familiar road, especially at night. I think my thoughts, I may sing along to the radio, but I'm also a careful driver: I watch the road, I notice the mile markers appear one by one in my headlights and recede in a stream in the rearview mirror. And yet when I reach my exit I'm surprised that I've done so and am unable to account for all the turns that brought me there. It's unsettling: was I not paying attention after all? Clearly I must have been, or I might not now be alive. So how did I get here? Where did the time go?

For that matter, when we say, "I lost track of the time," what we're typically saying is that we weren't tracking the time to begin with. Wearden conducted a study that bears this out. He gave two hundred undergraduates a questionnaire asking them to describe an occasion in which time seemed to have elapsed faster or slower than normal. He also asked them to describe in detail what they were doing at the time; to recall whether they noticed in the moment that time was moving faster or slower; and to note which drugs, if any, they were on. The students responded with statements such as

> Time flies when I'm out with friends either drinking or had some coke. Dancing, chatting. Next minute you know, it's 3 a.m.

> Alcohol consumption seems to lead to time speeding up—possibly due to the fact that I am socializing at the same time and therefore having fun.

Overall, Wearden found, students reported that the experience of time passing faster than normal was more common than the experience of time passing slowly. Distortions of either kind were two-thirds more likely to occur if the subject was somehow intoxicated; alcohol and cocaine seemed to contribute to time's flying, whereas marijuana and ecstasy seemed equally likely to make time speed up or slow down. Time consistently sped up when subjects were busy, happy, concentrating, or socializing (alcohol was often involved) and slowed down at work or when subjects were bored, tired, or sad. Strikingly, many said that they felt no sense of time flying until they were nudged by some external marker of real time—sunrise, a glance at a clock, the bartender's last call. Before that, they often had no sense of time at all. As one subject remarked, "I generally only become aware of the time when the bar/pub I am in begins to close or someone around me makes me aware of what time it is."

The reason that time flies, at least on the scale of minutes to hours, is so straightforward that it's almost circular: it flies because you aren't regularly looking at the clock. Afterward you notice that, say, two hours have gone by since you last thought about the time; you're aware that two hours is a fairly long time, but since you didn't tabulate and remember each minute of it, you infer from the large number of accompanying events that the time passed quickly. As one of Wearden's subjects put it, "After taking cocaine with 2 friends and sitting round her house after a night out which ended at approximately 3 a.m., it seemed to all of a sudden be 7 a.m., so therefore time had passed quicker than I thought it had."

It's no different from what we experience on waking up in the morning or, for that matter, when daydreaming. "Some chance idea fills the whole field of our consciousness," Paul Fraisse wrote in *The Psychology of Time*, "and when a clock chimes in the distance we are amazed that it is so late in the night or in the morning. We have not been conscious of a duration." Fraisse added that this also explains why many people find that monotonous tasks actually go by quickly: when you're bored you're thinking about time, maybe even looking at your watch, but when you're daydreaming you aren't. A 1952 study by Morris Viteles, an industrial psychologist at the University of Pennsylvania, found that only twenty-five percent of workers who engaged in seemingly monotonous tasks actually experienced them that way. (Among his many accomplishments, Viteles developed the Viteles Motorman Selection Test, to help the Milwaukee Electric Railway hire the best streetcar drivers; wrote *The Science of Work* and *Motivation and Morale in Industry*, and once gave a lecture titled "Machines and Monotony.")

Wearden also noticed that whether or not a stretch of time flies by depends on *when* you think about it—retrospectively or as the experience is unfolding. Time can crawl in either the past tense or the present tense; a traffic jam or a dinner party might be lasting an eternity while you're in them, and you'll likely remember them that way later. But time rarely seems to fly in the moment, Wearden said. That's virtually by definition: time flies because you aren't currently tracking it. What was the last movie you sat through thinking, "Wow, this movie is really flying by!" Either you're bored and glancing at your watch or you're immersed in the film and unaware of the time. At meetings and conferences, Wearden likes to ask fellow psychologists whether they've ever experienced time moving quickly or if they know anyone who has. The answer is always no.

"The consensus among psychologists, after a few beers, is that the experience of fast time is so rare as to be nonexistent," Wearden said. "You can't fast-forward time while you're still in it." Time doesn't fly when you're having fun: it's found to have flown only once the fun is over.

"Set the timer, Daddy!"

Joshua has wandered into the kitchen, where I'm making the morning coffee. He and Leo are two years old and, empowered by language, they brim with complaints about each other: he has the Thing, why can't I, it's not fair. Each wants to assert his nascent self, yet only perfect parity can make the universe right. Susan and I have instituted a turn-taking policy, but we're soon taught a basic lesson in temporal perception: to the boy without the Thing, the other boy's turn is always longer. Duration is very much in the eye of the beholder, not the holder.

So I've introduced a clock, one of those egg timers that you set by twisting it and then wait as it ticks down the seconds and a tinny bell rings. The boys like this; it's not an arbitrary judge, nor irritable and half-shaven, nor likely to be preoccupied with reading the news. Its objectivity seems almost magical and they regularly call on me to invoke it to settle their disputes. But even this strategy is wearing thin for them. Joshua has taken to grabbing the timer and twisting it to make it ring, again and again, as if this will end his brother's turn and compel him to hand over the Thing. If time bends, surely it can be bent to his will.

Typically I set the clock for two minutes, but one day Susan sets it for four minutes, to give us time for half a conversation. Midway through—eerily close to the two-minute mark—Joshua comes in, troubled: why hasn't the timer rung yet? Evidently, with regular training in the two-minute turn, he has learned the interval; I've succeeded in getting time into him. "It's as if they learn time the way they do a language," Susan says. She's right, to a degree that we as parents have yet to fully appreciate. But it's more complicated too. Our children clearly possess some sort of timer already—a nascent version of the clock that leaves me impatient at a stoplight or on the train platform, certain that my turn should have arrived by now. I can get time into my children but only insofar as they already have some means in place to capture it.

In 1932 Hudson Hoagland went to the drugstore. Hoagland was a respected physiologist in the Boston area, with a particular interest in how hormones affect the brain; in the course of his career he taught at Tufts Medical School, Boston University, and Harvard, and helped start a foundation that developed the birth-control pill. At one point, in the nineteen-twenties, he investigated a high-society medium named Margery who was eventually debunked by Houdini. At the moment, though, Hoagland was buying aspirin; his wife was at home with the flu and a temperature of 104°F and she had sent him out to the pharmacy.

The trip took all of twenty minutes, but when he returned his wife insisted that he'd been gone for far longer. Hoagland was intrigued. He asked her to count out sixty seconds while he timed her with a stopwatch; she was a musician, with a trained sense of how long a second ought to last, but she counted to sixty in just thirty-eight seconds. He repeated the experiment two dozen times over the next few days and found that, as she recovered and her temperature fell to normal, her counting slowed back to normal too. "She unknowingly counted faster at higher than at lower temperatures," Hoagland noted in a journal article some years later. When he repeated the experiment with subjects who had fevers or whose body temperatures he raised artificially, the results were similar. It was as though the subjects had an internal clock and heating it up made it tick faster; they didn't feel that time was flying, but they were inevitably surprised at the end of the experiment to learn that, according to the clock on the wall, less time had passed than they thought. "In a fever, all things being equal, we could come early for our appointments," Hoagland wrote.

Hoagland's findings prompted other researchers to embark on what John Wearden has described in one review paper as "some of serious psychology's most bizarre experimental manipulations." Volunteers were placed in heated rooms and given sweatsuits to don or special helmets that warmed their heads, then asked to tap out thirty-second intervals, or to adjust the pace of a metronome—to, say, four ticks per second—or to report when four or nine or thirteen minutes had passed. In one experiment, subjects took timing tests as they pedaled an exercise bicycle in a tank of water. In a 1966 journal article, Hoagland reviewed his original findings and some of the subsequent scholarship and offered a physiological explanation. "The human time sense is basically dependent upon the velocity of oxidative metabolism in some of the cells of the brain," he wrote.

Hoagland's explanation hasn't held up (it's not clear that even he knew exactly what he meant by it), but interest in the general subject has only intensified since. Of the many facets of time, by far the most studied is our grasp of duration: one's ability to estimate how long an interval of time—typically a short interval, ranging from perhaps a couple of seconds to a few minutes—lasted. This is the span of moment-to-moment experience; in it, we plan, estimate, and make decisions; we daydream and grow impatient or bored. If you grow restless at a stoplight or annoyed by your certainty that your sibling has had the Thing for slightly longer than is fair, you are navigating this span of time. Many of our social interactions unfold in these tiny windows and depend on a keen sense of interval timing. A genuine smile typically begins and ends more quickly than a forced one does; the difference in timing is subtle but is noticeable enough that an observer can often distinguish the real thing from a fake.

For well over a century researchers have recognized that we shape time as we move through it; it seems to speed or slow depending on whether you're happy, sad, angry, or anxious, filled with dread or anticipation, playing music or listening to it; a study in 1925 found that a speech seems to go by more quickly to the person who gives it than to a person who listens to it. When researchers discuss "time perception," typically the time in question is just a handful of seconds or minutes.

As it happens, a young child's ability to know when two minutes are up aligns him with much of the rest of the animal kingdom. In the nineteen-thirties, the Russian physiologist Ivan Pavlov revealed that dogs are masters of brief intervals. Pavlov is mainly remembered for proving that if a dog hears a bell while it eats, it can be trained to salivate at the sound of the bell alone, a reaction called a conditioned response. Pavlov showed that a dog can be conditioned as readily to a time interval as to a bell. If the dog is given food every thirty minutes, eventually it will begin salivating near the end of that thirty-minute interval, even if it doesn't receive food. The dog has internalized the timespan, somehow counts off the minutes, and anticipates a reward at the end of it. Dogs have human-seeming expectations that can be quantified and corralled.

Laboratory rats exhibit similar abilities. Suppose you train a rat in the following way: a light turns on to mark the start of the interval, and if the rat waits, say, ten minutes to press a lever it gets a food reward. Repeat this a few times. Next, turn on the light but don't give the rat food, no matter how often it presses the lever. The rat's response will remain consistent: it starts pressing the lever shortly before the ten-minute mark, presses most frequently at ten minutes, then shortly afterward gives up. Like a dog, the rat frames its expectation around the interval; it also knows to stop responding soon after the interval ends if its expectations go unrewarded. And its expectant behavior scales across different time intervals; in general, whether a rat is conditioned on an interval of five minutes, ten minutes, or thirty minutes, it starts and stops pressing the lever at a moment representing ten percent of the interval as a whole. Faced with a thirty-second interval, the animal starts pressing three seconds before the interval begins and stops three seconds after it ends; with a sixty-second interval, it starts pressing six seconds early. In 1977 John Gibbon, a mathematical physicist at Columbia University, codified this relationship in an influential paper that laid out what he called scalar expectancy theory. The theory, sometimes referred to as S.E.T. (and said aloud as "set"), was essentially a group of equations demonstrating that an animal's expectancy—its rate of response— increases as the end of the conditioned interval approaches and does so in proportion to the interval's total duration. Nowadays any effort to explain how animals can time intervals must account for this scale invariance.

A rat can perform other uncanny feats of timing. Placed in a maze that offers two paths to a piece of cheese, the animal quickly learns to take not only the shortest physical path but also the fastest. If two routes are equidistant, each with a temporary holding area—a six-minute wait in one versus a one-minute wait in the other—the rat soon learns to choose the path that will lead to the food in the least amount of time. The animal can discriminate between time intervals and intuit how long is a waste of its time.

Ducks, pigeons, rabbits, and even fish can do some version of the same thing. (Gibbon worked with starlings.) In 2006, biologists at the University of Edinburgh demonstrated that hummingbirds exhibit timing abilities in the wild. The researchers set out eight flower-shaped bird feeders filled with sugar water; four refilled every ten minutes and the other four refilled every twenty minutes. The hummingbirds—three males that had set up territory around the fake flowers—quickly learned the two refill rates and anticipated them. They visited the ten-minute feeders significantly sooner than the twenty-minute feeders, actively avoided the latter until twenty minutes was nearly up, and started visiting all the feeders just before each was due to refill. They also showed an uncanny ability to remember where the flowers were and which ones they'd probed most recently; they wasted little time on empty flowers. To forage efficiently among true wildflowers, the bird must memorize the whereabouts of a variety of flowers, learn their recharge rates (which vary over the course of the day), calculate an optimal path through them, and aim to arrive at each flower before—but not too long before—one's competitors. Even in a field of plenty, time is of the essence, and hummingbirds work to make the most of it.

Of course, optimizing one's time is something that humans do all the time, across seconds and minutes, sometimes consciously and sometimes not. If I run, can I catch the train that's about to pull away from the platform? Is this checkout line taking too long—and at what point should I move to that other one? Making such decisions is only possible if I have some way of measuring these brief intervals and comparing them to one another. It seems like a sophisticated behavior but clearly it's fundamental to the animal kingdom, and the fact that it can be done by creatures with brains no larger than peas strongly suggests that there's some sort of timing device in there, one that is both basic and ancient.

For much of the twentieth century the study of timing and time perception was divided roughly into two schools, each largely unaware of the other's relevance if not its existence. One, centered mainly in Europe, was concerned primarily with the existential experience of time and with translating philosophy into psychology. The German experimentalists of the nineteenth century, preoccupied with psychophysics, treated time as a real thing; Ernst Mach wondered if humans had distinct receptors, perhaps in the ears, that are attuned to it. In 1891, in an influential essay titled "On the Origin of the Idea of Time," the French philosopher Jean-Marie Guyau dismissed the objective view of time and proposed a very modern, and also very Augustinian, idea: that time is in the mind alone. "Time is not a condition, but rather a simple product of consciousness," he wrote. "It is not an a priori form that we impose on events. Time, as I see it, is nothing but a kind of systematic tendency, an organization of mental representations. And memory is nothing but the art of evoking and organizing these representations." Time, in short, is our system for keeping our memories straight.

Subsequent researchers lost interest in the alleged *Zeitsein*, or "time-sense," and instead began to probe and document the many ways in which time perception could be led astray. Drugs such as pentobarbital and nitrous oxide made subjects underestimate an interval of time whereas caffeine and amphetamines caused them to overestimate it. A high-pitched sound seems to last longer than a low-pitched sound of equal duration. "Filled" time feels shorter than "empty" time: twenty-six seconds spent solving anagrams or printing the alphabet in reverse seems to pass more quickly than twenty-six seconds spent resting and doing nothing. Piaget was the first of many scientists to study how children perceive time—and to demonstrate that temporal perception is something that our species grows into over time.

In 1963 the French psychologist Paul Fraisse summarized the previous century or more of temporal research, including his own investigation, in *The Psychology of Time*. In its encyclopedic sweep, the book codified what until then had been a disparate field of study; it was as influential in its realm as James's *Principles of Psychology*. "It had a huge influence on the topics that graduate students selected for their doctoral dissertations," the cognitive neuroscientist Warren Meck, of Duke University, told me. "That was in the good old days, when writing a book meant something, at least in the sciences."

Meanwhile, in the United States, a separate group of scientists, including a young Warren Meck, were approaching timing from another direction, at first unaware that they were doing so. Meck is now considered the elder statesman of interval-timing research, and in recent years he has tried to rally the field around a core set of ideas. "I'm trying to herd the cats," he told me.

Meck grew up on a farm in eastern Pennsylvania and likes to say that he is still a farmer, insofar as he has spent much of his career in the lab raising, managing, and experimenting on rats and mice. He spent the first two years of college at the local branch of Penn State, which was just across the highway from his high school, then transferred to U.C. San Diego, where he worked as a research assistant in an operant-conditioning lab studying pigeons. At the time, in the nineteen-seventies, animal learning and conditioning was still dominated by behaviorism, a school of thought, promulgated in the United States by B. F. Skinner that sought to understand how animals learn by closely controlling what they do in the laboratory. Cognition and social psychology were of little interest to these scientists, who were reluctant to view their animal subjects as much more than ambulatory machines. Pavlov had demonstrated that an animal's ability to learn different time intervals is central to the conditioning process, but behaviorists typically viewed interval timing as a means to an end, not an end worth studying in its own right.

As Meck remembers it, the lab at U.C.S.D. resembled a telephone operator's command room, with relay lines running this way and that. In most such labs, the technology was crude enough that all the boxes had to be controlled in lockstep with each other. Conditioning mostly involved training pigeons to choose between various delays to reinforcement: the bird might get grain if, after seeing a particular color on the response key, it waited twenty seconds before pecking the key. "Fixed intervals, variable intervals—we thought that the animals behaved as if they were little clocks," Meck said. His colleagues wanted to know what kinds of things they could get the animals to learn, "but I was always interested in what in the brain allowed them to do that. That's not a question a Skinnerian would likely ask."

Meck went on to Brown University, where he studied with Russell Church, a well-known experimental psychologist and frequent collaborator with John Gibbon, the originator of scalar expectancy theory. By this time Gibbon had turned his attention full-on to timing, wondering aloud what cognitive processes enabled his animals to discriminate one brief time interval from the next. In 1984 the three researchers published a seminal paper, "Scalar Timing in Memory," that expanded on Gibbon's 1977 paper and laid out an information-processing model to account for the animals' timing behavior.

What they proposed was a basic clock, akin to an hourglass or a water clock, that does two things: it emits pulses at a steady rate, with some sort of pacemaker, and it stores the number of ticks or pulses during an event being timed, for later reference. It ticks and it counts the ticks; it's a clock with memory. In some versions the clock has a third feature, a switch that dictates whether or not the pulses accumulate. When the interval to be learned begins, the switch closes, enabling pulses to build up; when the switch opens, the pulses stop accruing. The researchers referred to their model as scalar timing theory but it is more commonly known as the pacemaker-accumulator model or, sometimes, the information-processing model. Something similar had been proposed a decade earlier by an Oxford psychologist named Michel Treisman, who applied the idea to studies of human behavior, but it was rarely cited; the new version was the first application to animal learning and it caught on right away.

When we spoke, Meck made a point of emphasizing that Gibbon's original paper on scalar expectancy theory, in 1977, made no mention of clocks, stopwatches, or pacemakers, although many contemporary scientists think that it did. "It was mostly a set of closed mathematical equations" that predicted the timing of keypresses and pecks by rodents and pigeons, Meck said. The subsequent paper, which Meck described as "a cartoon version of S.E.T.," introduced layman's terms as an "intentional contrivance" in order to help make the theory "more generally accessible to a broader range of psychologists, i.e., those who are less mathematically inclined." Internally the coauthors referred to scalar timing theory as "the S.E.T. model for dummies." The behaviorist mindset was still so strong that when Meck and his colleagues initially put the word "clock" in their paper, the journal's editors insisted that they take it out.

"That paper was somewhat risky for us," Meck said. "'Clock' is a cognitive construct that no self-respecting Skinnerian would ever use; if you can't see it, you can't describe it. Treisman didn't annoy anyone by using the word 'clock,' whereas we annoyed a lot of people in the animal field."

The pacemaker-accumulator model quickly became popular among animal researchers—those studying timing, anyway—because it offered a conceptual mechanism, if not a physiological one, to explain some of the temporal relationships they'd been observing. For instance, studies involving rats on various drugs suggested that stimulants—cocaine, caffeine—caused rats to overestimate brief time intervals. That makes sense if one imagines that these drugs cause the pacemaker to tick faster: more ticks accumulate in the memory bin than normally would in the same interval, so when the system goes back to "count" how much time has accumulated, it overestimates the duration. Drugs like haloperidol and pimozide, which reduce the effectiveness of dopamine in the brain and in humans are used as antipsychotics, have the opposite effect, slowing the tick rate and causing the rats to underestimate time intervals.

Similar results were seen in human subjects who'd been given these or similar drugs: stimulants sped up the clock, causing people to over-estimate time intervals, while depressants led them to underestimate. And there was mounting evidence that medical disorders could disrupt the pacemaker clock as well. Patients with Parkinson's disease suffer from low levels of dopamine in the brain, and in cognitive tests they consistently underestimate brief intervals of time, which suggests that the reduction of dopamine slows down the internal clock.

The pacemaker-accumulator model also helped to explain the curi-ous fact that, in experiments, whether a time interval seems longer or shorter than normal depends on *how* the subject is asked to respond. For instance, suppose you're asked to judge the duration of an audio tone; you could give your estimate verbally ("I think that tone lasted five seconds") or by reproducing it, perhaps by tapping, counting out loud, or pushing a button for what you consider to be an equivalent duration. And suppose that before you hear the audio tone you ingest a small dose of a stimulant such as caffeine. If you estimate the dura-tion verbally, you'll likely say that the tone lasted longer than it actu-ally did—but if you push a button for what you think is an equivalent length, your response will be shorter than the actual event. Such is the complexity of our internal clock that, if it is pharmacologically sped up, you can either overshoot or undershoot the same interval depending on how you provide your answer.

The pacemaker-accumulator model can account for the paradox. Let's say that the audio tone you heard was actually fifteen seconds long. Sped up by caffeine, your internal clock ticks faster than usual, so more ticks than normal accumulate during that span—maybe your clock ticks sixty times in that fifteen seconds instead of the usual fifty. (I'm pulling these numbers out of a hat.) When the beep ends, you're asked to verbally estimate the interval. Your brain counts up the ticks, and because more ticks equate to more time, and because sixty is big-ger than fifty, you'll report that the beep lasted slightly longer than it actually did. Now, instead, suppose that you're asked to estimate the beep's duration by pressing a button for an equivalent length of time. Your clock is ticking faster on caffeine, so you'll get to fifty ticks (your brain's measure of fifteen seconds) more quickly than normal, so you'll stop pressing the button before fifteen seconds is actually up. Your verbal guess will overestimate but to an observer your actions will seem to underestimate.

The pacemaker-accumulator model soon spread beyond animal-research labs to the labs of scientists studying time perception in people. "Traditionally, human researchers don't pay much attention to the work of animal researchers, and vice versa," Meck said. "Animal researchers tend to be reductionists and control freaks. But timing was different; John Gibbon brought human and animal researchers together for the first time. When we introduced the information-processing model of S.E.T. at a conference, the human researchers loved it."

John Wearden, in England, was among them. When the 1984 paper appeared, he saw an opportunity to shift his study group from rats to people, and he is now among the more avid proponents of the pacemaker-accumulator model. In one of his more provocative experiments, Wearden showed subjects a visual stimulus or had them listen to an audio tone of varying length. Just beforehand, though, he played a five-second train of audio clicks, at a frequency of either five or twenty-five clicks per second, on the hunch that this would cause the subject's interval-timing clock to speed up. It did: afterward, when his subjects were asked to estimate the duration of the stimulus, those who'd heard the clicks first consistently overestimated how long the stimulus lasted.

Wearden then wondered: If your clock can be made to tick faster such that a span of time dilates, can you accomplish more in that added time? Does time merely seem to stretch or, in some substantive way, does it really stretch? "Suppose, reading as fast as you can, you can read sixty lines of text in sixty seconds," Wearden said. "Then, by giving you some flickers or click trains, I've made it seem that the sixty seconds last longer than they did. Can you now read more than sixty lines in a minute?"

Yes, you can, it turns out. In one experiment, Wearden had his subjects view a computer monitor on which four boxes were arrayed in a row. A cross would appear in one of the boxes and the subject had to push one of four keys corresponding to its correct location. Wearden found that their response times increased noticeably if they heard a series of clicks—a five-second burst, at either five or twenty-five clicks per second—at the start of the experiment. In a similar experiment, the subjects saw not a cross but an addition problem along with four possible answers; again, they picked the right answer more quickly if they first heard a train of clicks.

In addition to reacting faster, people also can learn more in that time, he found. In another experiment, he briefly showed subjects a bunch of letters arrayed in three rows, for half a second at most, then immediately asked them to recall as many letters as they could. Again, listening to clicks beforehand increased the number of letters they could correctly recall, slightly but significantly. (It also increased the false-alarm rate: their recall of letters that were never there.) Speeding up their clocks—increasing the tick rate—seemed to provide subjects more time in which to remember and process information.

It had long been noted that one's estimate of a duration can vary widely depending on your circumstances: your emotional state, what's going on around you, and the particular events you're observing and timing. "Our feeling of time harmonizes with different mental moods," William James wrote. In the past decade or so, scientists have discovered ever more interesting ways to slow or speed up the interval-timing clock based on the subject's frame of mine, the content of what they're experiencing, or both. If you briefly observe an image of a face on a computer monitor, your estimate of how long it lasted will depend on whether the face is elderly or young, more or less attractive, or the same age or ethnicity as you are. Photos of kittens and dark chocolates last longer on screen than equally brief images of scary spiders and blood sausages. Not long ago I came across a paper titled "Time Flies When We Read Taboo Words," in which the researchers tested the time-distorting properties of various sexually charged, off-color words. For the sake of academic decorum, though, the taboo words weren't included in the published paper; a note at the end said that I had to request them directly from the author. I did, and when the list arrived I learned that *fuck* and *asshole*, when viewed on a computer monitor, seem not to last as long as words such as *bicycle* and *zebra* even though their actual duration onscreen is the same.

One aspect of the pacemaker-accumulator model that Wearden likes most is that it mirrors a common experience: as an event or a duration stretches on, you have the feeling of time building up inside you. One can imagine the internal clock as a sort of digital watch, with numbers that increase roughly in proportion to the passage of time outside. A longer span of clock time equals more internal clicks; more internal clicks translate as a longer passage of clock time.

People can actually do arithmetic with time intervals. In one experiment, Wearden trained his subjects to recognize an interval ten seconds long, by playing a beep to mark the start of the interval and another to mark its end. He did this a couple of times to accustom the subject to a standard interval. Then he presented a new interval, anywhere between one and ten seconds long, again bounded by beeps, and asked the subject to estimate what fraction of the standard it represented. Was it half as long? A third? A tenth? (To keep his subjects from counting internally to measure the duration—cheating—Wearden had them carry out a minor task on the computer screen while they were listening to the interval.)

"When you ask people to do that, the blood drains from their face, they think it's impossible," Wearden said. But their estimates turn out to be surprisingly accurate: the smaller the fraction they hear, the smaller their estimate of the duration. "Their estimates are almost completely linear. When you're halfway through the interval objectively, you're halfway through it subjectively as well—which implies some kind of linear accumulation process." And there was very little disagreement between one subject and another; one person's one-tenth or one-third of an interval was another's as well. Wearden also found that people are good at adding intervals together. He had subjects listen to two or three durations of different lengths and asked them to combine them in their minds into one longer duration, then had them try to match the sum to longer durations that he played for them. "They could do that pretty well," he said. "Now how on earth can you do that if you don't have an orderly metric of time?"

On a recent Saturday morning, Susan and I slipped into the city to visit the Metropolitan Museum of Art, a place we hadn't been to as a couple since before the boys were born. The crowds hadn't yet descended and for an hour or so we wandered around and absorbed the cavernous hush of art. We separated for a bit, apart but together; while Susan roamed among the Manets and Van Goghs I slipped into a small side gallery, not much larger than a subway car, that held a series of glass cases with small bronze sculptures by Degas. There were a few busts, several horses in stride, and the figure of a woman stretching, a small bronze rising to her feet and curling her left arm upward as if waking from a long nap.

At the end of the gallery, in one long case, were two dozen ballerinas in various states of motion or rest. One dancer was examining the sole of her right foot; another was putting on her stocking; a third stood with her right leg forward and her hands behind her head. Arabesque decant—tilted forward on one foot, arms outstretched, like a child imitating an airplane. Arabesque devant—upright on left leg, right leg pointed forward, left arm curled overhead. Their motion was frozen yet still fluid; I felt as though I had wandered into a rehearsal and the dancers had paused just long enough for me to appreciate the mechanics of their grace. At one point a group of young men wandered through whom I also took to be dancers. Their instructor said, "Quick, which one are you right now?" and they each picked out a bronze to emulate—the man nearest me with his right leg forward and his hands on his hips, elbows winged backward. "I like that you picked that one, John," the instructor said.

Time flies when you're having fun. It can slow in moments of duress, during a car crash or fall from a roof, or distort under the influence of intoxicants, moving faster or slower depending on the agent. There are myriad lesser-known ways to bend time too, and scientists are discovering more all the time. For instance, consider these two sculptures by Degas:

They belong to the ballerina series I'd seen at the Met, demonstrating the dance positions across the range of exertion; the ballerina on the left is at rest and the ballerina on the right is executing a first arabesque *penché*. The sculptures (and the images of them) aren't moving, but the ballerinas depicted seem to be—and that, it turns out, is enough to alter your perception of time.

In a study published in 2011, Sylvie Droit-Volet, a neuropsychologist at Université Blaise Pascal, in Clermont-Ferrand, France, and three coauthors showed images of the two ballerinas to a group of volunteers. The experiment was what's known as a bisection task. First, on a computer screen, each subject was shown a neutral image lasting either 0.4 seconds or 1.6 seconds; through repeated showings, the subjects were trained to recognize those two intervals of time, to get a feel for what each is like. Then one or the other ballerina image appeared onscreen for some duration between those two intervals; after each viewing, the subject pressed a key to indicate whether the duration of the ballerina felt more like the short interval or the long one. The results were consistent: the ballerina *en arabesque*, the more dynamic of the two figures, seemed to last longer on screen than it actually did.

That makes a certain sense. Related studies have revealed a link between time perception and motion. A circle or triangle that moves quickly across your computer monitor will seem to last longer on screen than a stationary object does; the faster the shapes move, the bigger the distortion. But the Degas sculptures aren't moving—they merely suggest movement. Typically, duration distortions arise because of the way you perceive certain physical properties of the stimulus. If you observe a light that blinks every tenth of a second and simultaneously hear a series of beeps at a slightly slower rate—every fifteenth of a second, say—the light will seem to you to blink more slowly than it does, in time with the beep. That's a function of the way our neurons are wired; many temporal illusions are actually audiovisual illusions. But with Degas there's no time-altering property—no motion—to be perceived. That property is entirely manufactured by, and in, the viewer—reactivated in your memory, perhaps even reenacted. That simply viewing a Degas can bend time in this way suggests a great deal about how and why our internal clocks work as they do.

One of the richest veins in temporal-perception research is on the effect of emotion on cognition, and Droit-Volet has conducted a number of compelling studies that explore the relationship. In a recent series of experiments, she had subjects view a series of images of faces, each of which was neutral or expressed a basic emotion such as happiness or anger. Each image lasted onscreen for anywhere from half a second to a second and a half, and the viewer was asked to say whether the image lasted for a "short" or a "long" time. Consistently, viewers reported that happy faces seemed to last longer than neutral ones, and both angry and fearful faces seemed to last longer still. (The angry faces lasted even longer to three-year-old children, Droit-Volet found.)

The key ingredient seems to be a physiological response called arousal, which isn't what you might think. In experimental psychology, "arousal" refers to the degree to which the body is preparing itself to act in some manner. It's measured through heart rate and the skin's electrical conductivity; sometimes subjects are asked to rate their own arousal in comparison to images of faces or puppet figures. Arousal can be thought of as the physiological expression of one's emotions or, perhaps, as a precursor of physical action; in practice there may be little difference. By standard measures, anger is the most arousing emotion, for viewer and angry person alike, followed by fear, then happiness, then sadness. Arousal is thought to accelerate the pacemaker, causing more ticks than usual to accumulate in a given interval, thereby making emotionally laden images seem to last longer than others of equal duration. In Droit-Volet's study, sad faces were deemed to last longer than neutral faces but not to the same degree as happy ones did.

Physiologists and psychologists think of arousal as a primed physical state—not moving but poised to move. When we see movement, even implied movement in a static image, the thinking goes, we enact that movement internally. In a sense, arousal is a measure of your ability to put yourself in another person's shoes. Studies find that if you watch an action—someone's hand picking up a ball, say—the muscles in your hand become primed for action. The muscles don't move but their electrical conductivity rises as if they were prepared to do so, and your heart rate increases slightly too. Physiologically speaking, you are aroused. The same will occur if you merely observe a hand resting next to an object—presumably primed to pick it up—or even just a photograph of a hand holding the object.

A wealth of research suggests that this sort of thing goes on all the time in our daily lives. We mimic each other's faces and gestures, often unknowingly; various studies have found that subjects will imitate a facial expression even when, through laboratory trickery, they aren't aware that they're seeing a face. Two friends in conversation will correlate their movements much more than two strangers will—and a third-party observer can tell which pairs are friends just by watching their conversation on video. Marnix Naber, a psychologist at the University of Utrecht, conducted a study that had pairs of subjects competing against one another in a variant of the arcade game Whac-A-Mole. As the game progressed, the players increasingly (and unconsciously) synchronized their movements, even when doing so lowered their scores. This sort of mimicry seems to be an integral part of socializing, and a sensitivity to timing is essential to it; the meaning of a nod, smile, or sigh can change dramatically depending on whether it's short or long, quick or slow, regular or sporadic.

Social mimicry also induces physiological arousal and seems to open a path that helps us perceive emotions in others. Studies find that if you make a face as though you're expecting a shock, the actual shock, when it comes, will feel more painful. Exaggerating your facial expression while viewing pleasant or unpleasant film clips amplifies your heart rate and skin conductivity, the typical measures of physiological arousal. Studies using fMRI have found that the same areas of the brain are activated whether the subject experiences a particular emotion, such as anger, or observes a facial expression of it. Arousal signals a bridge to the interior lives of others. If you see a friend feeling angry, you don't merely infer how she feels: you literally feel what she feels. Her state of mind, and state of motion, become yours too.

And so does her sense of time, it turns out. In the last few years, Droit-Volet and others have demonstrated that when we embody another person's action or emotion, we embody the temporal distortions that come with it. In one experiment, Droit-Volet had her subjects view a series of faces—some elderly, some young—briefly on a computer screen, in no particular order or pattern. She found that the viewers consistently underestimated the duration of the elderly faces but not the young ones. In other words, when the viewers saw an elderly face, their internal clocks slowed down as if to "embody the slow movements of elderly people," Droit-Volet writes. A slower clock ticks less often in a given interval of time; fewer ticks accumulate, so the interval is judged to be briefer than it actually is. Perceiving or remembering an elderly person induces the viewer to reenact, or simulate, their bodily states, namely their slow movement. "By means of this embodiment," Droit-Volet writes, "our internal clock adapts to the speed of movement of elderly people and makes the elapsed stimulus duration feel shorter."

Or recall Droit-Volet's earlier experiment, in which participants re-ported that angry and happy faces seemed to last longer onscreen than neutral ones. She had attributed the effect to arousal but she began to suspect that embodiment might play a role too. Perhaps the subjects were mimicking the faces as they viewed them and the act of imitation was giving rise to the time distortion. So she ran the experiment again, with a critical difference: one group of participants was asked to view the faces while holding a pen between their lips, to suppress their facial expressions. Viewers without the pen significantly overestimated the duration of angry faces and moderately overestimated happy ones—but the other viewers, their lips and faces constrained, detected virtually no temporal difference between the emotional and the neutral faces. Time had been righted by, of all things, a pen.

This all leads to a strange and provocative conclusion: the percep-tion of time is contagious. As we converse with and consider one an-other, we step in and out of one other's experience, including the other's perceptions (or what we imagine to be another's perception, based on our own experience) of duration. Not only does duration bend, we are continuously sharing these small flexions among us like a currency or social glue. "The effectiveness of social interaction is determined by our capacity to synchronize our activity with that of the individual with whom we are dealing," Droit-Volet writes. "In other words, individuals adopt other people's rhythms and incorporate other people's time."

Our shared temporal distortions can be thought of as manifestations of empathy; after all, to embody another's time is to place oneself in his or her skin. We imitate each other's gestures and emotions—but we're more likely to do so, studies find, with people with whom we identify or whose company we would like to share. Droit-Volet found this to be true in her study of faces: viewers perceived elderly faces as lasting less long onscreen than young ones, but only when the viewer and the viewed belonged to the same gender. If a man viewed an elderly female face, or a woman viewed an elderly male face, there was no temporal distortion. Studies of ethnic faces show a similar effect: subjects over-estimate the duration of the angry faces compared to neutral ones, but the effect is more likely and pronounced when both the viewer and the face are of the same ethnicity. Droit-Volet found that the viewers most likely to overestimate the duration of angry faces were those who measured highest on a standard test for empathy.

We step out of ourselves and into one another all the time, but we enter these exchanges with inanimate objects too—faces and hands, pictures of faces and hands, and other figurative objects, such as the balletic sculptures of Degas. Droit-Volet and her coauthors of the Degas paper argue that the reason the more dynamic sculpture appears to last longer on screen—the reason it's physiologically arousing in the first place—is that "it involved the embodied simulation of a more effortful and arousing movement." Presumably this is what Degas had in mind all along: an invitation to participate, an inducement to even the most lead-footed observer to step inside. I see a sculpture of a ballerina bending forward on one foot and in a small, outwardly imperceptible yet essential way, I am right there with her, performing my own internal arabesque. I am bronzed with grace and, in the moment of my gaze, time bends around me.

Emotional faces, moving bodies, athletic sculptures—all of these can induce temporal distortions, and in a manner that can be explained with the pacemaker-accumulator model. Yet they're also puzzling. Clearly, life dictates that we possess some sort of internal mechanism to keep time and monitor brief durations—yet the one we carry around can be thrown off course by the least emotional breeze. What's the point of owning such a fallible clock?

"The thing that strikes me about subjective time is how bad we are in comparison to a stopwatch," Dan Lloyd, a philosopher at Trinity College and a coeditor of *Subjective Time: The Philosophy, Psychology, and Neuroscience of Temporality,* told me. "We're inconsistent in all sorts of aspects and subject to all sorts of manipulations. It's a mystery to me that we function as well as we do."

Maybe there's another way to think about it, Droit-Volet suggests. It's not that our clock doesn't run well; on the contrary, it's superb at adapting to the ever-changing social and emotional environment that we navigate every day. The time that I perceive in social settings isn't solely mine, nor is there just one cast to it, which is part of what gives our social interactions their shading. "There is thus no unique, homogeneous time but instead multiple experiences of time," Droit-Volet writes in one paper. "Our temporal distortions directly reflect the way our brain and body adapt to these multiple times." She quotes the philosopher Henri Bergson: "*On doit mettre de côte le temps unique, seuls comptent les temps multiples, ceux de l'expérience.*" We must put aside the idea of a single time, all that counts are the multiple times that make up experience.

Our slightest social exchanges—our glances, our smiles and frowns—gain potency from our ability to synchronize them among ourselves, Droit-Volet notes. We bend time to make time with one another, and the many temporal distortions we experience are indicators of empathy; the better able I am to envisage myself in your body and your state of mind, and you in mine, the better we can each recognize a threat, an ally, a friend, or someone in need. But empathy is a fairly sophisticated trait, a mark of emotional adulthood; it takes learning and time. As children grow and develop empathy, they gain a better sense of how to navigate the social world. Put another way, it may be that a critical aspect of growing up is learning how to bend our time in step with others. We may be born alone, but childhood ends with a synchrony of clocks, as we lend ourselves fully to the contagion of time.

S ometimes when Matthew Matell gives a talk about his research, he begins by showing the audience a slide. It shows a printed sentence, which he reads aloud:

Interval timing is so entrenched in our moment-to-moment perception that it may be difficult to imagine what our conscious experience would be like without temporal expectation.

Halfway through, right after he says "it may be difficult," he stops abruptly and lets several increasingly awkward seconds pass. The audience shifts uneasily—*What's going on? Does he have stage fright?*—until he at last resumes. "I did that when I applied for my position here at Villanova," Matell told me. "Afterward, my sponsor came up and said he thought I'd completely frozen up, it made him so anxious."

But the audience reaction proves his point: we are so closely attuned to time's passage from moment to moment that we barely give it a thought until our expectations are violated. "You weren't timing my interval," he said. "But when it's interrupted, you're suddenly aware that you were timing the whole time." Early on, academic advisers tried to steer him away from studying time: why bother with such an esoteric subject? "But that's not seeing the forest for the trees," he told me. "Timing is so embedded in everything we do, it's impossible to imagine experience without it."

Matell is a behavioral neuroscientist at Villanova University, outside Philadelphia. Often when he tells someone new that his research explores how we perceive time, they press on him the usual questions: Why do I wake up at the same time every day even without an alarm clock? Why am I always so tired in the middle of the afternoon? Those are questions for a circadian biologist. Matell studies interval timing, the mechanism governing the brain's ability to plan, estimate, and make decisions over periods lasting from roughly one second to several minutes.

But what is the nature of that mechanism? Does the brain have a central interval timer analogous to the master circadian clock in the suprachiasmatic nucleus? Is there a distributed network of clocks that kick in according to the task at hand? For thirty years the pacemaker-accumulator model has served as a reliable platform for experiments on time perception; it's clear that our duration judgments can be manipulated as easily and predictably as our judgments about brightness or sounds. But the model is and always has been a heuristic device, the sort of clock one draws on a napkin; where in our three-pound collective of neurons is it actually? "It exists conceptually," Wearden said to me at one point. "It exists mathematically, as a framework for stimulating research and explaining research. But whether there's a physical mechanism that does this sort of thing remains to be seen."

For some psychologists the answer holds little interest. In the preface to *The Psychology of Time Perception*, Wearden writes that "none of the topics treated in the book would be significantly illuminated in any way by the neuroscience of timing in its present state, at least in my view." Neuroscientists beg to differ. People who suffer from certain real-world disorders, including Parkinson's disease, Huntington's, schizophrenia, and even autism, are known to have difficulty with timing tasks. Interval timing clearly has a biological basis, and a better understanding of it could shed light on these disabilities, or at the very least shed further light on the workings of the human brain. Something makes us tick—what is it? That's what Matell, among others, wants to find out.

I found Matell's office in the top corner of an older building on the Villanova campus, up four flights of marble steps worn round with age. Classes had just let out for the summer, and the linoleum-tiled corridors were deserted. The quiet made everything seem bigger than normal, and I began to think that I was back in my elementary school or tracing a path down some other recess of memory. After a left turn, the hallway narrowed and, after a few doorways, appeared to end. I asked around and learned that what looked like an exit door opened onto a cul-de-sac with a warren of offices and lab rooms.

Matell appeared, dressed in a T-shirt, shorts, and hiking sneakers, and greeted me energetically. He was on his way to a part of the lab he called the rat room, and he wore a pair of stretchy blue gloves; years of handling rats had made his skin allergic, and the grad student who typically managed the rats was out that day. Matell spoke rapidly but warmly, and his eyes widened as he talked. At one point he said, "Science is about making up stories and seeing if they hold any water."

For its first century or so, the study of time perception mostly documented the cognitive manifestations: how a subject, human or nonhuman, responds after being presented with a stimulus (bright flash, angry faces, Degas sculptures), and under what conditions that response could be altered (cocaine, a fall from a hundred-foot height, riding a bike in a tank of water). But increasingly researchers are able to ask where and how the brain produces such responses. Microtargeted drugs can turn off or amplify select clusters of neurons to gauge their role in time perception. Brain-imaging tools reveal which groups of neurons are called upon when a subject engages in timing tasks. The psychology of time has given rise to the neuroscience of time. As Matell and other researchers venture inside our heads, they confront the essential human mystery: how does a three-pound mass of cells generate the memories, thoughts, and feelings that we associate with ourselves? How does the wetware give rise to software? One researcher told me that we are all neuroscientists insofar as we all know equally little about how the human brain gives rise to the human mind.

"The brain functions like a corporation," Matell said. "There are lots of units doing what they do, maybe some top-down management. But each unit is doing its own thing, and within each unit are individuals"—he was referring to neurons—"each doing their own job. I tend to make the analogy of neurons to people. They're little information-processing assemblies. At some level, neurons are acting like automatons. The big question is, how do physiological systems, like brains made of neurons, give rise to psychological phenomena like consciousness? We like to think we have free will, but I don't think you can be a neuroscientist and truly believe that. That suggests that our behavior is operated by something other than our brain."

The human brain is an assembly of some hundred billion neurons. A neuron is like a living wire; it transmits information, in the form of an electrochemical pulse, from one end of its extended cell body to the other, mostly in one direction. Some neurons are long—the sciatic nerve, which runs from the base of your spine to your big toe, is about three feet long—but most are microscopic, and all are exceptionally thin; packed side by side, anywhere from ten to fifty could fit within the width of the period that ends this sentence. Each has a receiving end, composed of branching dendrites, which under a microscope look like the roots of a tree; a long cell body, or axon, along which the signal propagates; and a branched terminus through which the signal is passed on. A typical neuron receives input from ten thousand or so "upstream" neurons and transmits to a smaller set of neurons downstream. Neurons usually aren't connected to one another physically; they communicate across tiny gaps, or synapses. When a signal reaches the terminus of one neuron, it triggers the release of neurotransmitters that travel across the synapse and attach to the dendrites of nearby neurons, like keys fitting into a set of locks. If the signals reaching one neuron are sufficiently strong, they prompt the neuron to generate its own signal to pass along. A neuron either fires or it doesn't, and once it does, its action potential is always the same; all that can change is its rate of fire. A stronger stimulus—a bright light—will induce a neuron to fire more often than a weak stimulus will, so that neuron is more likely to trigger neurons downstream. Even at the scale of cells, time—input per unit time—plays a part.

Neuroscientists sometimes describe neurons as "coincidence detectors." A neuron is always receiving some baseline drip of input from upstream; only when the drip becomes a torrent, and a large number of signals arrive simultaneously, is the neuron prompted to fire. One might fairly ask what "simultaneously" means at this scale—what is "now" to a neuron? A brain cell works something like a water clock. The neurotransmitters from upstream attach to its cell membrane and open channels that let in ions—usually sodium ions, which have a slightly positive charge. These start to depolarize the cell, and when depolarization reaches a critical threshold, the neuron fires; the faster the input, the faster the tide of ions rises. But it's a water clock with holes: the ions leak out through the cell membrane and the cell can actively pump out more. "The whole thing can probably be modeled with a leaky wineglass with a fragile stem and some Manischewitz," one researcher told me. "Pour enough in fast enough and the stem will break; otherwise you just dribble wine on your tablecloth."

"Now" is however long it takes for the incoming tide of ions to outpace the exiting flow. It is a dynamic window of time and very much under the control of the cell. The neuron can pump ions out quickly or slowly, and the number of ion channels on the cell membrane is regulated by the cell's DNA. The neuron also gives differential weight to the input from upstream: a signal that arrives from a neuron farther out on the dendrites degrades more on its way to the axon, and so may factor less in whether the neuron fires. "I think of neurons as individuals that are computing something," Matell said. "They're integrating information—action potentials—over time and space." By way of an analogy, Matell said, he asks his students a question: on a Saturday night, how do you decide whether to go to that frat party or stay in and study? "You weigh your sources," Matell said. "If you ask your mom, she'll say one thing; ask your friends, they'll say another. Then again, maybe your friends think you should go, but you've been to other parties that those friends have suggested and had a terrible time, so their opinion means less."

In any case, "now" to a neuron is not zero. Here, as anywhere else, it takes time to make time: fifty microseconds (one-twentieth of a millisecond, or one-twenty-thousandth of a second) for neurotransmitters to diffuse across a synapse from one neuron to the next; maybe twenty milliseconds for a neuron to depolarize in advance of firing; another ten milliseconds or so for its own signal to travel the length of the cell.

A neuron can fire ten to twenty times a second, and when groups of neurons periodically fire in unison, and do so regularly, their pulses register as electromagnetic oscillations. "One of the challenges of understanding time perception is that the brain's processes are operating on a millisecond timescale," Matell said. How does the same circuitry give rise to our ability to navigate seconds, minutes, and even hours? One early model focused on the cerebellum and treated it almost literally as an electrical circuit, with branching networks and delay lines that might slow down a signal. That concept helps to explain some behavior, such as our ability to determine the direction of a sound. (When an auditory signal reaches one ear slightly before the other, the lag provides information about the location of the sound.) But it's less applicable to the perception of intervals that are seconds to minutes long. Instead, for the past several years, Matell has been helping to explore another model, one that works less like a telephone circuit and more like a symphony.

In 1995, having graduated from Ohio State, Matell went to Duke to pursue his doctorate. He studied under the cognitive neuroscientist Warren Meck, who had arrived the previous year from Columbia, intent on trying to understand the neural basis of interval timing. By now Meck had compiled two enlightening sets of data. One, derived from studies done in rats and people, revealed that one's sense of duration could be sped up or slowed down by administering drugs that altered dopamine levels in the brain. The second focused on circuitry. Research with rats indicated that if a part of the brain called the dorsal striatum was destroyed or removed, the animal lost the ability to carry out standard timing tasks. And there was mounting evidence—by Chara Malapani, at Columbia, but since bolstered by work from several researchers, including Marjan Jahanshahi, a neuroscientist at University College London, and Deborah Harrington, at U.C. San Diego—that patients with Parkinson's disease, who suffer from damage to the striatum, also consistently misjudge time intervals. Soon after Matell arrived, Meck handed him both data sets.

"He gave me these papers and said, 'Your job is to figure out how all this works in the brain,' " Matell told me. "I don't think he meant, like, come up with the answer. But I started reading a lot of papers from the neurobiology literature rather than the psychology literature."

As he spoke, Matell showed me around his lab and the setup for his rats. Each rodent occupied a plastic chamber roughly a cubic foot in size. Each chamber had a small speaker that would play the occasional audio tone, a way to deliver food pellets, and three holes that the rat could stick its snout into. "Holes work better than levers, because rats like to poke their noses into stuff," Matell said. With this configuration he could train the rats to learn time intervals of his choosing. For instance, if the rat poked its nose into a hole (an action that was detected by an infrared beam across each hole) it would be rewarded thirty seconds later with a food pellet. If it was impatient and poked before the thirty seconds elapsed, nothing happened; success, on the rat's part, thus required both poking and waiting—and learning how long to wait until it poked again. In 2007 researchers at Georgia State University found that chimpanzees were better at waiting out a thirty-second delay to get a candy treat if they were able to distract themselves in the meantime—by playing with toys and flipping through copies of National Geographic and Entertainment Weekly that the researchers had given them. Matell's rats spent the time grooming and sniffing around. "If they were humans, they'd probably get on their phones and surf the Internet," Matell said.

Once the animal learns a particular interval, Matell can try to disrupt that knowledge. In some experiments he might give the rat a drug—perhaps a specific dose of amphetamine, microinjected into a particular part of the brain—to see how that speeds or slows the animal's timing, to begin to decipher which neural structures are involved. Or he might selectively damage or destroy a particular organ within its brain, to measure how the animal's timing is altered. The procedure is delicate and can be imprecise; typically the target is a tiny region in the brainstem called the substantia nigra pars impacta, which in rats is no larger than a BB. "Just like humans, rat brains aren't identical," Matell said. "You're basically shooting in the dark a little bit." He showed me an oversize book called the Atlas of Brain Maps. Every page displayed a successive slice of a rat's brain, measurable to the millimeter; it looked like a *Gray's Anatomy* of cauliflower. When an experiment ends, Matell said, the animal is euthanized and its brain is removed and finely sliced, and the slices are mounted on slides and examined in comparison to the images in the book. "So we can say, 'We aimed for this structure—where did we end up?'"

Another way to study how a rat learns to time intervals is to implant electrodes in its brain and record the neural activity as the animal carries out its timing tasks. This too is delicate work. Matell showed me what looked like an inch-long, fibrous sword: a small metal platform, like a hilt, from which eight short wires protruded, each with an electrode at the end. With the brain atlas as a guide, Matell or a grad student carefully inserts the electrodes into the rat's brain; the wires are attached to a cable that runs up through the top of the experimental chamber and to a recording device, so the rat can move around in its box relatively unhindered. Whatever neural spikes occur are time-coded and can later be compared to the rat's activities.

"It's like taking a microphone and putting it in a room full of people," Matell said. "Those people are neurons. You can listen to different things. Neurons have different voices, depending on the cell size or distance from the electrode."

At one point, Matell stopped at a metal cabinet and took out a plastic model of a human brain. He set it on a table and began to pull it apart, separating the right hemisphere of the cerebral cortex from the left. Inside, sitting atop the brainstem, was a structure that looked like a flat toadstool; this was the corpus callosum, the bundle of nerve fibers that provides a critical connection between the two hemispheres. Matell pointed out a wishbone-shaped structure embedded in each hemisphere—the ventricles, fluid-filled sacs that, among other things, provide internal cushioning. "The brain is in fluid and surrounded by fluid," he said. "It's like an egg-protection system." Beneath the corpus callosum was the hippocampus and amygdala, which are part of the limbic system, the seat of emotion and memory, as well as the thalamus, the basal ganglia, and other subcortical structures.

As a species that thinks, we're accustomed to thinking that the brain's main job is to help us think. The brain is essential to that task, but ultimately what it does is help us to anticipate, to move, and to select the best movement for whatever situation the body faces in the moment. Achieving this goal requires the brain to minimize uncertainty about which moves to make; and achieving *that* goal demands that it first gather solid data about what's happening out there and, in particular, how everything's going so far—what the results of the previous moves were and whether the situation is getting better or worse. Toward that end, information travels through the brain in a sort of loop-the-loop. Sensory data come in—through the eyes or ears or spinal cord—then pass through distinct areas of the thalamus before radiating out to the sensory cortical areas: the primary visual cortex, in the occipital lobe, at the back of the brain; the primary auditory cortex, in the temporal lobes, on either side; and the somatosensory cortex, in the parietal lobe, toward the back of the head. From here the streams combine and are routed down into the limbic system and frontal lobe. This is sometimes called the What Pathway, through which the brain figures out what a stimulus is, devoid of any value. Is it cake or is it a snake? Once that assessment has been made, the information travels into the limbic system, including the amygdala and hippocampus, where it is encoded with value (how badly do I want that cake?) and, if it's worth remembering, recorded. The data then moves on to the frontal cortex, where decisions are weighed (should I eat the cake before I do my homework, or wait till afterward?), priorities are set, and less pertinent information (my diet) can be downgraded. From there, it's on to the premotor and motor areas—located at the top of the brain, next to the sensory areas—which initiate action.

Located roughly midway on this trip is an important region called the basal ganglia, a conglomerate of structures that includes the striatum and the substantia nigra pars impacta; signals enter through the striatum, which in textbook illustrations has a spiral shape and looks something like a telephone earpiece. The basal ganglia are the brain's labor-saving department. If my typical response to a piece of cake is to eat it immediately, my brain soon figures out that it can skip the usual looping What Pathway—see cake, identify cake, recognize cake as desirable, debate whether to eat cake—and get right to the business of eating. By recognizing particular patterns of firing among the cortical neurons, the basal ganglia get me what I want faster while freeing up my neural architecture for new stimuli. The basal ganglia are where rote activities are learned and habits, even addictions, are formed.

It's also a central component of the brain's interval timing clock, Matell and Meck believe. Every neuron in the cortex is a bit like an antenna. "It's tuned to some particular thing," Matell said. "It's a state-of-the-world detector for some restricted state." The cortex in turn sends thousands of neurons into the basal ganglia, which is composed of hundreds of thousands of spiny striatal neurons; every striatal neuron monitors the state of ten to thirty thousand cortical neurons, with plenty of overlap, so each one is good at detecting particular patterns of firing that occur upstream. When a particular pattern occurs, the striatal neuron fires, triggering neurons in the nearby substantia nigra pars impacta to release dopamine—a little neurochemical reward that helps mark the pattern as memorable and worth noting in the future. The signal goes on to the thalamus, the motor neurons, and back to the cortex. "The contribution of all those inputs is what the basal ganglia striata are detecting," Matell said. "It's like a habit-learning center. In rats it's involved in timing, because timing is a behavior that the rat has learned and that's now rote."

Those mechanics are well established. The model, Matell said, proposes that interval timing is made possible by the fact that, when stimulated by external signals, groups of cortical neurons fire with distinctive patterns. Some display what are called theta oscillations, firing at a rate of between five to eight times per second, or five to eight hertz. Others oscillate at eight to twelve hertz—alpha frequencies—and still others at twenty to eighty hertz, or gamma oscillations. These oscillations are detected in turn by the spiny neurons in the dorsal striatum. Of course, Matell said, these firing rates are well below the timescale that we encounter in our daily, conscious lives. "The brain operates on a millisecond scale, yet we can time things up to a few hours. You've been here, what, an hour and a half? We can estimate that without looking at a clock. So how do we go from millisecond operations in the brain to minutes-to-hours scale of operation?"

To address that conundrum, Matell and Meck borrowed from a model developed by another neuroscientist, Chris Miall, of the University of Birmingham. We headed back to Matell's office as he continued to explain. Large windows let in the bright sun of late spring and a view across the campus rooftops. On one wall was a tall bookshelf with titles such as *Psychopharmacology* and *The Wet Mind*, and on a nearby windowsill I noticed an unopened novelty toy called The Incredible Growing Brain, which required only the addition of water. On another wall was a whiteboard; Matell found a marker and began to draw on it.

He drew two rows of hashmarks, each representing the firing rate of a neuron. Now suppose, he said, that some stimulus begins, such as an auditory tone. Your neurons begin firing right away and continue for the duration of the tone, but they don't all fire at the same rate; maybe one spikes every ten milliseconds and the other spikes more often, every six milliseconds. Now suppose the two neurons are wired into the same striatal neuron, which detects when the two neurons fire simultaneously; this happens every thirty milliseconds.

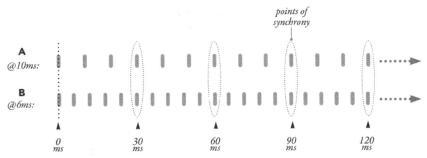

Neuron A fires every ten milliseconds and neuron B fires every six milliseconds.

The result, Matell said, is a striatal neuron capable of detecting an interval—thirty milliseconds—that is considerably longer than either cortical neuron generates by itself. And each striatal neuron has thirty thousand cortical neurons plugged into it, not just two; it might detect the coincident firing of dozens or thousands of neurons at once. With that math in play, the spiny neurons in the basal ganglia could be attuned to a wide range of real-world time intervals well beyond the millisecond timescale.

In fact, it may be that virtually every possible duration is being realized by one's neurons at every moment—it's just that the brain doesn't bother to remember them all. Learning a particular duration is simply a matter of reinforcement: the rat gets a food pellet; the human earns candy, verbal encouragement, or some other positive reward. (Wait at the traffic light for ninety seconds, until it turns green, then you win the satisfaction of being free to go.) The reward prompts a burst of dopamine to be released into the basal ganglia, and the pattern of cortical firings is duly noted, sent on to the thalamus, and retained in memory for later reference.

Strictly mathematically, when one considers the billions of neurons in the brain and the billions upon billions of signals they exchange moment after moment, it seems almost inevitable that they would generate some way to time events in the outside world. And yet it's marvelously inconceivable—that rote interactions among living cells could give rise to computation and to a behavior as intimate and instinctive as the ability to judge the duration of a stranger's smile. A world of typing monkeys should have a better chance of reproducing one of Shakespeare's plays.

When I spoke to Meck, he emphasized that what he, Matell, and similar researchers are trying to shed light on isn't "timing as normally defined" but temporal discrimination—the process of learning that a certain duration has more value than another durations. "The brain is timing things all the time even if you're not paying attention to them," he said. "Ten seconds wouldn't mean anything to you if we didn't tell you it was important. You're learning to discriminate what's good and what's bad; something matters. And to discriminate, you need memory. I don't know of any timing task that isn't a temporal discrimination."

Matell and Meck refer to their model of interval timing as the striatal beat-frequency model, and they describe it in musical terms. The basal ganglia is the conductor, its spiny neurons continuously monitoring the cortex for groups of neurons firing in synchrony; in one paper, Meck and Matell refer to this as "the composition of cortical activity." (Scientists who study time seem to be fond of musical analogies.) "It's like an orchestra whose playing simultaneously tells me where I am in the task," Matell said. I asked him what he meant. He reminded me that the basal ganglia is critical to the formation of habits—behaviors that we act upon based on our surroundings without realizing that we're doing so. Much of driving, he said, is "automatic, habit-based processing." You see a certain exit sign and know to put your blinker on, then move into the right lane, then position your hands a certain way on the steering wheel to turn it.

"The cortex detects the exit sign, it triggers the striatum, the striatum recognizes this activity in the cortex that's essentially this real-world pattern and says, Okay, make this behavioral shift, which is to put on the blinker," Matell said. "That movement is detected by the cortex and triggers another behavioral shift, which is to move into the right lane. The blinker is detected, which sets off another behavioral shift, which is to slow down the car. And you just keep going through these sets of chaining behaviors: I detect this particular environment, I do this particular behavior, which puts me in a new environment, et cetera."

Learned durations arise from these same loops of data and, at least initially, are closely tied to actual tasks. A rat waiting for its food pellet is like a person at the symphony. "It's not that the animal knows where it is in time, it just knows that food is coming," Matell said. "I don't have a sense of time passing, just a behavior that occurs." He added, "Let's say you've listened to a symphony a hundred times. Now you start something on the stove, you put some water on to boil, and then you go and you turn on the symphony, and you know that when you get to the third measure of the second movement, that's when your water is going to boil. You recognize that certain measure of the second movement by the amalgamation of what you're hearing. You're not recognizing it because it's louder than it was at the beginning of the symphony. Nothing's growing: it's not becoming more complex; there's nothing systematically changing in magnitude. It's different than the pacemaker model and this feeling of accumulating or depleting. Food comes at brain state ten instead of brain state thirty and I'm probabilistically going to do my thing."

He recalled an incident from his days as a grad student. He and his wife were watching a movie and paused the videotape to go into the kitchen. Back then, hitting pause didn't quite stop the tape but instead sent it into a microloop—moving forward a quarter of a second and then backtracking, replaying the same instant over and over. After five minutes or so, the tape would resume on its own. Matell and his wife had been in the kitchen for some time when something seemed amiss. "We were both like, hmm, shouldn't the pause have stopped?" he said. "Neither of us were aware of doing this; we weren't attending to time because we were busy fixing food. But we both were suddenly struck by the fact that something didn't happen that should have happened. Temporally, something wasn't the way it should be. That seemed very consistent with this sort of pattern detection—oh-we're-in-the-third-measure-of-the-second-movement kind of feeling—without any buildup to it."

Matell is quick to emphasize that whatever the neural basis for timing may be, it's not the same as having an organ for sensing time. The ears detect sound waves, the eyes detect light waves, the nose interprets molecules. "Unlike with other sensory systems, there's no 'time material' that we have a detector for," Matell said. "Clearly the brain does sense time and controls our behavior, but what the brain is measuring is not objective. It's subjective time. The brain is paying attention to its own functioning in order to derive some temporal landscape." As far as human perception is concerned, time is the brain listening to itself talk.

The striatal beat-frequency model is gradually gaining a foothold in the neuroscience literature; increasingly it's cited by other scientists and is seen by many as the leading neurophysiological explanation for interval timing. But the book on timing isn't entirely closed. Often when the striatal beat-frequency model is mentioned in a paper, it's accompanied by a qualifying statement, such as "hardly any convincing evidence exists that would show how specific physiological cycles function as an internal clock for judging time," or a note suggesting that scientists "have so far been unable to identify a simple neural mechanism dedicated to the processing of time." Other models are being kicked around. "There must be ten new computational models of timing ability that come out every year," Patrick Simen, a neuroscientist at Oberlin College, told me. Simen and his colleagues introduced their own in 2011, the opponent process drift-diffusion model, which borrows components of an established decision-making model and also invokes the coincidence-detection abilities of the basal ganglia. "In some sense this model might be kind of new, but it's taking off-the-shelf models and putting them together in slightly different way," Simen said. Warren Meck had said virtually the same thing about his own model. "We're not coming up with new ideas per se," he told me. "I take pleasure in that. It's the I.B.M. model: we just took off-the-shelf components and assembled them in a way to make them more useful."

Even Matell has hesitations about the striatal beat-frequency model. For one thing, he said, it requires individual cortical neurons to show oscillations, which they generally don't do. "That may or may not be a problem," he said. Maybe a neuron fires in concert with a particular oscillatory signature but not all the time. Maybe the firings are like the yard markers on a football field, he said: there underneath it all, but hard to see after a muddy game. The model is also highly sensitive to noise—the small, constant variations in performance that exist in all biological systems. "That's all well and good if all the neurons are noisy together," Matell said. "But if you have something where one oscillator got a little faster but another got a little slower, then the model completely falls apart and it can't time it all. It becomes very, very sensitive to everything being coherent, and I don't think real life is like that."

And, like many scientists, he is haunted by the seemingly metric nature of time perception—the feeling that time "grows" or the ability to sense when one is, say, halfway through a given interval. It's an experience that even the rats in his lab have. Matell trained a group of rats under two conditions: when he sounded a tone, they would expect food after ten seconds; and when he turned on a light, they would expect food after twenty seconds. But to his surprise, when he turned on both stimuli, the rats expected food at fifteen seconds—midway between the two stimuli, as if they were averaging out the durations.

"I'm very convinced that animals have a perception of time that has some magnitude components," he told me. "They're not only averaging the durations, they're averaging the durations and weighing the result by the likelihood that each cue will pay off. They're behaving in a manner that suggests that some aspect of their information processing capacity has very quantitative, analog-like information. I still kind of buy the general framework that we set up: the striatum is sitting there looking at ensembles of cortical neurons, and when food comes, it releases a dopamine pulse that causes the striatal neurons to stamp in the ensemble of cortical neurons that are coactive, and then you're sitting there looking at the striatal neurons waiting for those coactive events to occur. But the patterns of activity across the cortex are not growing.

"And this is where I'd say the field is, and what the puzzle is: what's the pattern of cortical activity that allows time to flow in a manner such that, psychologically, it feels like it's growing? Is there any way in which we could have a pattern-recognition model that leads to behavior that is more ordinal? I'm fairly convinced that that's likely to occur, that there's gotta be some amalgamation of those two ideas. But I currently don't know how to get there.

"I'm not convinced that there's any one answer to the problem. So I hate to leave you with this scenario, like I have no idea what is going on. But really I don't. I don't have a firm grasp on how the brain is doing this right now." For a more upbeat response, he said, I should talk to Meck. "Maybe he's just been around longer than I have, and he doesn't have the same self-defeating attitude that I do. So he's more willing to promote the model, whereas I'm always happy to point out all the problems with it."

I called Meck a few days later. "It's a terrible clock from a physicist's point of view," he conceded; it shows tremendous variability, its ensembles of neurons may drift as far as ten to twenty percent out of sync. By comparison, he noted, the circadian clock has a variability of only one percent "but it has little flexibility—it can only measure twenty-four hours!" His clock, in contrast, is tremendously flexible, capable of timing a range of seconds to minutes while still exhibiting scalar invariance, and it helps to explain the temporal disorders experienced by people with Parkinson's and schizophrenia. It doesn't depart from scalar timing theory but builds on it, complete with a clock module and a memory module, "making it more 'biologically plausible'—that's the phrase we prefer to use," he said.

"Look," he said finally. "It's important to people that I not overturn the pacemaker-accumulator model just for the sake of doing something new; it has importance as a heuristic model in cognitive psychology. If your work doesn't have a need for you to go beyond that, you can stay with it. But for me, to be an academic, you're an explorer and you want to see how things work, particularly in the brain. I see my mission as hanging around the field long enough, and being sufficiently dogmatic and professional, to beat down all these other crazy ideas out there about modality differences, multiple timescales, memory decay, and the rest. It takes a long time."

Meck is ready for the field to move on. He entered it at a time when the very notion of an internal clock was anathema to behavioral biologists. The next step was to decipher the physiology; that's an ongoing endeavor, but the underlying premise—that there's some sort of timing mechanism, or mechanisms, in there to be explored—is no longer in question. "We studied timing to the exclusion of everything," Meck told me, describing the first generation of time researchers. "We tried to strip down the tasks so that all you're looking at was timing." The current generation, he added, "is looking at stuff more in the real world. They wouldn't claim that timing is so special—it's just part of what the brain does when it's learning, or attending, or experiencing emotions."

Catherine Jones, a cognitive neuropsychologist at Cardiff University, agreed. "My understanding of timing has evolved a lot," she said. "When I went in, in the late nineteen-nineties, the problem was set up already, the question of this internal clock being located somewhere in the brain. It was a bit of a silo. The thinking has broadened a bit. Now, when other people mention something, I think, Oh, that's related to timing—for instance, how we coordinate our speaking and gestures to make us better communicators."

Jones's first research post was in the lab of Marjan Jahanshahi, at University College London, looking at motor and timing deficiencies in Parkinson's patients. She now studies autism and wonders whether some of the behaviors common to the disorder—repetitive movements, difficulty with social interactions, difficulty integrating input from the different senses—could also be thought of as disorders of timing. Melissa Allman, a young behavioral and cognitive neuroscientist at the Michigan State University who has collaborated with both Meck and John Wearden, is pursuing a similar line of research. "I became interested in whether these behaviors might be explainable if you think of someone with autism as kind of lost in time," she told me. She and Jones both emphasized that this line of inquiry is still new and speculative; there's no specific theory or even an agreed-upon set of temporal difficulties associated with autism. But one day, they said, it might be possible to identify some deficiency in timing that shows up early in infancy and could serve as a screening test for children at risk.

Annett Schirmer, a psychologist at the National University of Singapore, started out studying emotions and nonverbal communication but was drawn into timing after marrying Trevor Penney, one of Meck's former graduate students. "I'm now part of the timing mafia," she told me. Schirmer noted that most of the studies on emotional arousal and timing have involved visual stimuli; for instance, it's well established that images of angry faces seem to last longer onscreen than neutral stimuli of equivalent durations. But in her own work she has found that aural stimuli have the opposite effect: the word *ah*, expressed with surprise, seems to listeners to have a shorter duration than a neutral *ah*. It's not clear why, Schirmer said, although sounds and voices introduce additional, dynamic variables, including tempo, that are absent in static imagery. In any case, the idea that arousal distorts time by speeding up the internal clock may not be so clear-cut.

"It's a viable mechanism," Schirmer said. "But there are probably other mechanisms by which our perception is influenced." One is attention. In the timing literature, attention is typically described as having the opposite effect as emotional arousal. Angry faces seem to last longer than neutral ones because they're arousing, which causes the internal clock to speed up, whereas taboo words, when viewed on a screen, seem to last shorter than neutral ones because they grab your attention; the brain is distracted from counting ticks, loses track of a few, and ends up underestimating intervals. But it can be hard to distinguish the two categories; on the face of it, words such as *fuck* and *asshole* would seem to be just as likely to arouse as to grab your attention.

"That's the tricky thing," Schirmer said. "Much of the evidence for the arousal model could be interpreted as attention. Maybe arousal *is* attention—that's a possibility. From a functional point of view the two are very tightly linked. From an evolutionary point of view, the things that are critical for survival typically capture our attention and are behaviorally arousing. For something to stand out, it needs to stand out in time so that we act on it and remember it."

If anything, timing research runs the risk of spreading itself too thin. "I think time is a much bigger landscape than any one researcher is able to cover—I don't think you can," Jones told me. "Where is the taxonomy of time?" A "taxonomy of time" is a time researcher's cry for help—a wish for some sort of overarching scheme that will bring order and consistency to a sprawling field of study. The phrase has been popping up more often in the literature lately, most recently in a 2016 paper that Meck coauthored with Richard Ivry, a psychologist and neuroscientist at the University of California at Berkeley. "A modern 'taxonomy of time' is very much needed," they write. "Researchers coming from different disciplines tend to invoke different terminology, use different experimental approaches, and sometimes focus on distinct questions within a specific context. As the field matures, it may be advantageous to find a common language to better articulate the questions that are being posed."

A common language. I find myself thinking back to my meeting with Felicitas Arias, the director of the Time Department at the Bureau International des Poids et Mesures, just outside Paris, when she showed me the world's most accurate clock: a sheaf of papers stapled in the corner—now an email blast—that is shared universally. This is how we all agree to be on the same time. Timing researchers need something similar, maybe a new journal or two: *Timing & Time Perception,* or *Timing & Time Perception Reviews,* or one of several others that have begun publishing. What they need is the linguistic version of a clock.

T he next time I spoke to John Wearden, a couple of years had gone by. He had basically retired, he said—but a moment later he added that he'd found retirement "rather boring" and had started teaching again. He had a few studies going but mostly was helping younger colleagues with their research. His mother had died, at ninety-one. He had traveled to Egypt and South Korea and bought himself a "retirement car"—a Porsche that sounded a warning if he went faster than eighty miles an hour.

Certain aspects of time perception still vexed him, however, among them the age-old question of why time seems to go by faster as you get older. Of all the puzzles that time presents, this one may be the most common, most intimate, and most confounding. In studies—and there have been several—as many as eighty percent of subjects say that time seems to have sped up as they've aged. "*The same space of time seems shorter as we grow older*—that is, the days, the months, and the years do so," William James wrote, in *Principles of Psychology*. "Whether the hours do so is doubtful, and the minutes and seconds to all appearance remain about the same." But does time really fly as we get older? As ever, the answer very much depends on what you mean by "time."

"It's a very tricky question," Wearden told me. "What on earth do people mean when they say time goes more quickly? What's the right thing to measure? Just because somebody says they think that time is going quickly, or they agree when you ask, Does time go more quickly as you get older?—'Oh, yeah, it definitely does'—doesn't mean they're right. People assent to all sorts of things. It's an unexplored question, really. And we haven't begun to use the right tools experimentally, or in terms of recording what happens in real life, to get a handle on it."

There are at least two ways to phrase the time-and-age puzzle. Most often what's expressed is something like the following: a given span of time seems to pass faster now than it did when you were younger. A year, say, seems to go by faster when you're forty than it did when you were ten or twenty. James cited Paul Janet, a philosopher at the Sorbonne: "Whoever counts many lustra in his memory need only question himself to find that the last of these, the past five years, have sped much more quickly than the preceding periods of equal amount. Let any one remember his last eight or ten school years: it is the space of a century. Compare with them the last eight or ten years of life: it is the space of an hour."

To account for the impression, Janet offered a formula: the apparent length of a given span of time varies inversely in proportion to your age. One year seems five times shorter to a fifty-year-old man than to a ten-year-old boy, because a year is one-fiftieth of the man's life and only one-tenth of the boy's. Janet's proposition spawned a series of similar explanations for why time seems to speed up with age; call them ratio theories. In 1975 Robert Lemlich, a retired professor of chemical engineering at the University of Cincinnati, added a twist to Janet's formula. (Lemlich was perhaps better known as one of the inventors of an industrial process called foam fractionation, which uses a flowing foam to remove contaminants from a liquid.) Lemlich proposed that the subjective length of a span of time varies inversely to the square root of your age. He wrote out an actual equation,

$$dS_1/dS_2 = \sqrt{R_2/R_1}$$

in which dS_1/dS_2 is the relative speed with which a time interval seems to be passing compared to some years ago; R_2 is your current age, and R_1 is your age back then. If you're forty, a year seems to go by twice as fast as it did when you were ten, since the square root of 40 ÷ 10 is 2. (Lemlich was careful to note that his formulation "assumes the absence of any extended traumatic or unusual experiences.") The implications of his equation can be dispiriting. Strictly speaking, if you're forty and have a life expectancy of seventy, you've lived fifty-seven percent of your life, but by Lemlich's math you've lived $\sqrt{(40/70)}$, or seventy-five percent, of your subjective total life. (On the upside, according to Lemlich's math, you'll never feel like you have half as much time remaining in your life as you actually do.)

To test his equation, Lemlich ran an experiment. He gathered thirty-one engineering students (with an average age of twenty) and adults (average age forty-four) and asked them to estimate how much faster or slower time seemed to pass now compared to two periods in their lives: when they were half their current age and when they were a quarter their current age. Nearly all replied that time moved faster now than it had at both of those earlier points. A few years later, James Walker, a psychologist at Brandon University, in Manitoba, got similar results when he asked a group of older students (average age twenty-nine) "how long a year appears at present" compared to how long they'd thought a year lasted when they were one-half and one-quarter their current age. Seventy-four percent reported that time had passed more slowly at a younger age. Between 1983 and 1991, Charles Joubert, a psychologist at the University of North Alabama, ran three more comparable studies which also seemed to confirm Janet and Lemlich.

The problem with phrasing the research question this way is that it takes an impossibly optimistic view of human memory. I can't remember what I had for lunch last Wednesday, much less whether it was better or worse than lunch the Wednesday before. How likely am I to accurately recall a far more abstract experience—the speed with which time seemed to pass—from ten, twenty, or forty years ago? Moreover, as even James noted, ratio theories don't explain much: Janet's formulation "roughly expresses the phenomenon," he wrote, but "can hardly be said to diminish the mystery." James thought it more likely that the time-speeds-up-with-age experience results from the "simplification of the backward-glancing view." When we're young virtually every experience is new, so it remains vivid years later. But as we age, habit and routine become the norm; novel experiences are fewer (we've done everything already) and we barely take note of the time we currently inhabit. Eventually, James wrote, "the days and the weeks smooth themselves out in recollection to countless units, and the years grow hollow and collapse."

James's glum proposition belongs in the category of what one might call memory theories, along the lines of what Locke proposed: we judge the duration of a past span of time by the number of events that we remember having occurred in it. A period busy with memorable events will seem in retrospect to have passed slowly—to have taken more time—whereas an uneventful stretch will seem to have sped by, leaving you to ask where the time went. There are several potential ways that memory could influence the speed at which time seems to have gone by. Emotional events tend to loom large in memory, so, to an overworked parent, your four years in high school—the first prom, your first car, graduation, all highlighted in memory with the help of photos and scrapbooks—may well seem to have lasted longer than the average four-year-period or longer anyway than the past four years of your current life of commuting, errand running, and dishwashing. We also seem to remember certain periods of life, typically our teens and twenties, more vividly than others—a phenomenon called the reminiscence bump, which may contribute to the feeling that a given stretch of time lasted longer back then.

Baked into memory-based explanations is the assumption that as we grow older our lives become comparatively less memorable. But there's little evidence to indicate that it's true, and common experience seems to contradict it. The night that I met my wife stands out far more clearly in my memory than my first kiss at summer camp. I don't remember the weather or how old I was when I first rode a bike, but I do recall the bright spring Saturday a few years ago, when, at age forty-six, I let go of the seat of a six-year-old's bike and watched him careen ahead of me for the first time under his own wobbly power across the grass of a baseball field. In five decades I've traveled, loved, lost, and started over, but increasingly it feels as if the memories from the early years belong to someone else or to past lives and that everything notable that's happened to me has occurred in the years since I was married and became a parent. In that time, two boys have taken shape before my eyes, and everything new to them has felt new to me too, twice over: the alphabet, addition, multiplication, the piano, the Four Questions, and the ability,

after much practice in the backyard and with one's left foot, to softly bend a soccer ball into the top right corner of the net.

Time does seem to have sped up—it's going fast, certainly—but what do I mean by that? That less has happened to me in recent years than ever before? Or that I've come to identify with my kids' experience of time, which seems less burdened and hurried than my own and therefore makes mine feel that much more pressed? It can't be that my time is flying by because its contents are *less* memorable, so maybe it's the opposite: they're more memorable, or the memorable events are more numerous, making me more acutely aware of all the potentially memorable things I'd like to do but don't have time for and never will. Has time sped up as I've aged? Or does its constant speed simply bother me more because I have less time in front of me, so it feels more precious?

One of the earliest attempts to untangle these threads, even before Lemlich's, was a 1961 study titled "On Age and the Subjective Speed of Time," and it's a good lesson in bad science. The researchers noted that one thing that seems to make time go by faster is the feeling of being busy. "Is it being busy, itself, that is the important factor," they asked, "or is it that being busy may make time more valuable?" They gathered two groups of subjects: a hundred and eighteen college students and a hundred and sixty adults between the ages of sixty-six and seventy-five. Each individual was given a list of twenty-five metaphors to consider:

a galloping horseman
a fleeing thief
a fast-moving shuttle
a speeding train
a whirligig
a devouring monster
a bird in flight
a spaceship in flight
a dashing waterfall
a winding spool ,
marching feet

a large revolving wheel
a tedious song
wind-driven sand
an old woman spinning
a burning candle
a string of beads
budding leaves
an old man with a staff
drifting clouds
a stairway leading upward
a vast expanse of sky
a road leading over a hill
a quiet, motionless ocean
the Rock of Gibraltar

The subjects were instructed to consider how appropriate each met-aphor was "in evoking for you a satisfying image of time" by labeling the five most appropriate with a 1, the next five most appropriate with a 2, down to the five least appropriate and so on. The results suggested that the young and the old experienced time similarly; both groups felt that the most representative metaphors were ones such as "a fast-moving shuttle" and "a galloping horseman," whereas the least representative were phrases such as "a quiet, motionless ocean" and "the Rock of Gibraltar." But after some additional—and, to a modern reader, suspiciously convoluted—statistical stirring, the researchers concluded that older adults tend to consider swift metaphors more representative of their experience than static ones, while young adults generally preferred static metaphors.

However, the study also included a glaring methodological flaw. The authors had set out to determine which factor contributes more to the impression that time is passing quickly: how busy you are, or how much you value your time. If busyness matters more, they reasoned, then it should be young people who report that time speeds up, because young people are more active than old people. But it's the

older people who say it; therefore, the researchers conclude, the value of one's time is the larger factor because "time is running out for the older person as death approaches." However, beyond stating that "the aged individual is less busy and active than before," the authors never bothered to demonstrate that that's actually so. And the only measure of how people value their time was how they ranked the temporal metaphors. Like so many attempts to explain why time seems to speed up as we age, the study is little more than a supposition wrapped in numbers.

There's yet another, simpler explanation for the mystery of why time seems to speed up as we get older: it doesn't. Granted, that's a given— time doesn't *actually* speed up with age, that's just an impression. But a number of researchers have come around to thinking that the impression itself is illusory. Time only appears to appear to quicken as we age.

At first glance, the many previous studies seem consistent in their results: more than two-thirds of subjects—from sixty-seven percent to eighty-two percent—report that time seemed to pass more slowly when they were younger. But if the impression is to be taken at face value, one would expect to it emerge progressively with age. If, on average, a year seems to go by faster when you're forty than when you were twenty, then surveys should find that more forty-year-olds than twenty-year-olds say that time is going faster than before. Or if you asked both groups to characterize how fast the past year went by, the forty-year-olds would say that it passed more quickly. Some sort of gradient would be apparent, with the time-flying impression becoming more pronounced among older respondents.

But the numbers don't show that. Consistently, the impression is shared equally across age groups: two-thirds of older people say that time passes more quickly now than it did when they were younger— and so do two-thirds of young people. In equal proportions across all ages, people say that time has sped up with age. The result is a paradox: most everyone of all ages has the impression that time passes more quickly with age, which suggests that the impression, if that's what it is, has little to do with age.

So what's going on? Clearly a lot of people are experiencing *something*—what is it? Part of the confusion stems from the way that these studies ask their subjects to think about time. In one way or another, all of them ask a question that can't be reliably answered: how did you experience the passage of time ten, or twenty, or thirty years ago? Instead, if there's anything to measure, it's how a person feels about the passage of time right now. Here the ground is slightly firmer. In general, the impression that time is speeding by correlates more strongly with a person's psychological state, especially with how busy he describes himself as being, than with his age. As Simone de Beauvoir put it, "The way in which we experience the day-to-day flow of time depends on what it holds."

In 1991, Steve Baum, a psychologist at the Sunnybrook Medical Center, in Toronto, and two colleagues took a closer look at busyness and time perception in the geriatric crowd. They interviewed three hundred older people, mostly retired Jewish women, between the ages of sixty-two and ninety-four; half were active and the others less so, and many of the latter group lived in institutions or facilities for the aged. The subjects were first asked a series of questions meant to gauge their emotional health and happiness. Then they were asked, "How rapidly does time seem to pass for you now?" and instructed to respond with a 1 ("faster"), 2 ("about the same"), or 3 ("slower"). No particular time interval was specified—a week, a year—and it was left vague what "faster" or "slower" alluded to. (Faster or slower than what or when?) Still, the results were in line with other studies: sixty percent of the subjects said that time passed more quickly now than it had before. But in addition, the individuals who said this tended to be more active than their peers, led what they described as purposeful lives, and said they felt younger than their chronological age. Thirteen percent of subjects actually said that time moved more slowly now—and these individuals were more likely than the others to exhibit signs of depression. "Time does not speed up for persons as we age," the researchers concluded. Rather, they wrote, it speeds up with one's psychological well-being.

The strongest evidence against the notion that time seems to speed up with age comes from a trio of studies conducted over the past decade or so. In 2005 Marc Wittmann and Sandra Lehnhoff, of Ludwig Maximilian University of Munich, asked some five hundred German and Austrian subjects, ranging in age from fourteen to ninety-four and divided into eight age groups, a series of questions such as

How fast does time usually pass for you?
How fast do you expect the next hour to pass?
How fast did the previous week pass for you?
How fast did the previous month pass for you?
How fast did the previous year pass for you?
How fast did the previous ten years pass for you?

To each one, the subjects were instructed to respond on a five-point scale, from "very slowly" (−2) to "very fast" (+2). Again unlike earlier studies, this one didn't bother to ask people how they feel about a time interval now compared to their impression of it at some earlier point in their lives. Instead it asked people of different ages how they feel now about the speed of various time intervals; it's all in the present tense.

The results were quite clear: for each of the time intervals, each age group responded, on average, with a 1 ("fast"); there was no statistical difference across the age groups and little indication that more older people than younger people felt that time moved faster. Only one category showed a very slight difference: older subjects were more likely than younger ones to say that the past ten years had passed quickly. But the effect was small and seemed to top out at around age fifty; from age fifty to age ninety-plus, everyone responded that the past decade had passed at about the same fast ("1") pace.

A very similar experiment, conducted in 2010 with more than seventeen hundred Dutch subjects ranging in age from sixteen to eighty, found virtually the same thing. Again, every age group responded, on average, that each time interval, from one week to ten years, went by "fast" ("1"). The researchers, William Friedman, of Oberlin College, and Steve Janssen, of Duke University and the University of Amsterdam, found no statistical difference across the age groups and little indication that more older people than younger people felt that time moved at a fast ("1") pace. The only blip, as in the Wittmann and Lehnhoff study, was a slight indication that as people became older, they became more likely to say that the past ten years had passed quickly—at least up until age fifty, at which point the response leveled off.

What little variation Friedman and Janssen saw in the responses was attributable not to age but to how much time pressure the subjects currently perceived themselves as being under in their lives. In addition to the questions about time's passage, Janssen and Friedman had posed a series of statements designed to gauge the subjects' sense of busyness, such as "There is often not enough time to do everything I want or need to do" and "I frequently have to rush to make sure everything gets done." The subjects responded on a scale from –3 ("strongly disagree") to +3 ("strongly agree"), and the results mapped closely to their perception of time: those individuals who reported that the hours, weeks, and years were going by "fast" or "very fast" were also more likely to report that their lives felt busy or that they were unable to accomplish everything they wanted to do in a given day. In 2014 the researchers repeated the study with more than eight hundred Japanese subjects of all ages, with essentially identical results. All told, it seems, time speeds up not with age but with time pressure, which explains why people of all ages say it's accelerating: time is the one thing that virtually everyone in equal measure feels he or she lacks. "Everybody feels that time moves quickly on all scales," Janssen told me.

Still, there's that intriguing blip: in the Janssen and Friedman studies, as in Wittmann's, more older people than younger ones reported that time had accelerated over the past ten years. The past decade went by slightly faster for people in their thirties than for people in their twenties, and slightly faster still for people in their forties (all within the range of "1," or "fast"). But for everyone fifty and older the pace of the past decade stayed pretty much the same. Janssen is still sorting through the possible explanations, but he thinks it probably isn't related to time pressure: people are pretty good at assessing how much time pressure they've been under for the past week, month, or year, but probably not for the past ten years. (Besides, once the average subject reaches age thirty, it's a safe bet that they've been quite busy for the preceding ten years—as busy as the average fifty-year-old.) Maybe younger people can look forward to bigger life events and this anticipation makes their most recent ten years seem to have passed slowly. Maybe people in their twenties and thirties remember more events from the past ten years than older people do, making that decade feel relatively extended. But if that's what it is—if decades seem to speed up with age because our later years have fewer memorable events—why doesn't the effect continue to grow instead of leveling off after age fifty?

There's one more plausible explanation for why people older than fifty are slightly more likely than younger adults to say that the past decade has passed more quickly, Janssen and Wittmann think. It's the power of suggestion: the impression that time flies by more quickly as we age is a folk belief, one that older people are more likely than younger people to be influenced by when they evaluate the pace of the past decade. Consider again the evidence. The impression is held across age groups, widely and uniformly—despite the fact that it doesn't accord with most people's actual experience. A forty- or fifty-year-old is no more likely to say that the past year—or week, or month—went by "fast" than a twenty-year-old is. That's because what we're experiencing is related less to age than to how equally busy we all feel across the shorter time intervals. But when assessing the speed of the past decade, people older than fifty are willing to let some other consideration come into play—some factor that clearly doesn't increase in potency as their age climbs into the eighties and nineties. That factor, the researchers think, is the common notion that time flies by more quickly as we age, and older people are more likely to think that it's shaping their perspective.

This explanation is unnervingly circular: time seems to speed up as we get older because other people say it does. But I can also see how it might apply. For the longest time I ignored or dismissed the adage that time flies as we get older because I didn't feel old enough for the "as we get older" clause to apply. Lately, though, I've started to think that I am, and that it does. Time isn't speeding up; it's pace is cruelly steady, a fact of which I am ever more painfully aware.

One day, on an errand, I took the subway to Grand Central Station, in midtown Manhattan. The subway platforms are deep in the station; stairs lead up to a pedestrian level, where commuters come and go through turnstiles, and an escalator goes up to the mezzanine. At the foot of the escalator a middle-aged woman was handing out pamphlets. She wore a yellow T-shirt with the words "The End" on it, and the front of the pamphlet that she gave me also read "The End." She was shouting: "God's coming, we all know that! How can we be ready if we don't know the date?"

At the top of the escalator a man—older, with glasses, and slightly stooped—was also handing out pamphlets. He too wore a yellow shirt that read "The End," and just under those words was a date, May 21. That day was less than three weeks away. My immediate thought was uncharitable: on May 22nd, when it was clear that the world hadn't ended, what would they do with the leftover T-shirts? But I soon looped back around to the mortal proposition. What if everything really did come to an end next month—or next week, for that matter, or in the next few minutes? The end might be a cataclysm; it might be an aneurysm; it might be an anvil falling on me from ten floors up. I could die in my sleep. Was I ready? Had I made the most of my time? Was I doing so at that very moment?

In 1922 the Paris newspaper *L'Intransigeant* posed a question to readers: if you knew that the world was about to end catastrophically, how would you spend your last hours? Many readers responded, among them Marcel Proust, who took delight in the question. "I think that life would suddenly seem wonderful to us if we were threatened to die as you say," he wrote. "Just think of how many projects, travels, love affairs, studies, it—our life—hides from us, made invisible by our laziness which, certain of a future, delays them incessantly." His point being, how unfortunate that it takes being aware of an ending to focus one's attention on the present. Much of what we do in the present is done reflexively; habit is the enemy of mindful thought. Why don't we think more about the present when we're in it?

Recently, looking back through my journal, I discovered an entry from some years ago when I was passing through Grand Central on my way to the library to return a copy of Heidegger's *The Concept of Time*. The book, published in 1924, is essentially a bound lecture and it sketches out many of the ideas that appear later in Heidegger's opus, *Being and Time*. I'd kept it for many weeks until I'd discovered that it was due back that very day, so I'd found myself on the train to New York trying to reabsorb Heidegger's ideas about time in a rapidly dwindling window of my own.

Central to his argument is an amorphous concept that he calls *Dasein*, which translates literally as "there-being" or "being there" but which he also defines as "being-in-the-world," and as "being-with-one-another," and as "that entity in its Being which we know as human life," and even as "being questionable." (My own feeling is that if you have to cook up another word in order to define time, you aren't helping much.) The most concrete thing that Heidegger can say about *Dasein* is that it can't be fully defined until its end—after which, of course, it no longer is. "Prior to this end, it never authentically is what it can be."

Heidegger started out as a theology student (he later joined the Nazi Party) and was a close reader of Augustine, who had explored a somewhat similar idea. Begin by sounding a note or a syllable; its duration can't be measured—is it long? is it short?—until the sound ceases. *Now* can't be gauged until later, in retrospect. Heidegger extends the analogy to being in general: one's existence can't be fully evaluated until it's over. A question such as, "Am I making the most of my time?"—whether applied to the next hour or to the whole of one's time on earth—can't be answered without acknowledging that the time will end. Existentially, time gains its value from its finitude; *now* is defined by *later*. "The fundamental phenomenon of time is the future," Heidegger writes.

The rub, of course, is that in Heidegger's scheme the existential question can never find a satisfying answer; by the time you can supply one, you'll be dead. Augustine proposed that time might be nothing other than "the tension of consciousness"—the present mind straining between memory and expectation. Heidegger offers a more fraught tension, in which we are forever straining toward the future to evaluate our present life as if in hindsight. One's being—*Dasein*—is always "running ahead to its past," and this very act is the definition of time. Simply reading Heidegger's passages is enough to induce anxiety: "Dasein, conceived in its most extreme possibility of Being, *is time itself*, not *in* time. . . . Maintaining myself alongside my past in running ahead I have time."

I did not have time. When my train arrived at Grand Central, I rushed through the station, beneath the vault of painted stars, past the information booth with its globular clock, and went down into the subway to head to the library, having made a few hurried notes for my future self that I hoped to decipher later.

Right around the time that Joshua and Leo turned four, the hard questions began: What is "die"? Will you die? When will you die? Will I die? Are people made of meat? Do people decay? When I die, who will blow out my birthday candles—and who will eat my cake?

I wasn't entirely unprepared. The developmental psychologist Katherine Nelson has noted that the self begins to gel at around this age. For the first couple of years of life a child doesn't discern between her own memories and ones that are recounted to her; tell her about your trip to the supermarket and she'll likely remember the event as if she went herself. The experience of recollection is itself so new that it's as if all memories belong to her. Gradually she recognizes her memories as hers alone and so becomes aware of her continuity and passage through time: I am me, an awareness contained in a membrane, made up of my memories (I was me yesterday) and my expectations (I will be me tomorrow); I was and always will be me.

This stage of development was crystallized for me at breakfast one morning when one of the boys described a dream he'd had the night before—the first dream he could actually remember upon waking. It was a nightmare, he said: he was walking in the dark when an invisible voice confronted him and asked, "Who are you?" It was clear to me, if not to him, that the voice was his own. So here were two selves confronting each other—one self unknown to itself—at least one of which was self-aware enough to ask humankind's most existential question. But once a new self realizes its continuity across time, it pauses. *I will always be me*—how long is *always*? A self capable of noticing that everything around it expires can't avoid concluding that it will too, somehow, sometime.

The boys shared a bedroom, and at bedtime I would sit between their beds, with the lights out, and tell a story. One night, before the story had begun, I noticed that one of the boys was quietly crying. I asked what was wrong, and he said, "What happens at the end of the world?"

"I don't think anybody knows," I said.

"But what if I live *past* the end of world?" He was sobbing now. His concern, as I teased it out of him, was not that he would die one day but that he wouldn't—that he would be left utterly alone. Before I could find something to say that might be remotely reassuring and not factually incorrect, his brother chimed in.

"That's impossible," he said, and added, "If I'm lucky enough, I'll probably live to a hundred and three years old. Maybe even a hundred and fifteen."

The first boy stopped crying. "You can't go past a hundred and twenty," he said. He'd recently been reading the *Guinness Book of World Records*.

"Pretty much not, no," I said. "But nobody knows when they're going to die."

"It's all about how much you exercise," the brother said.

"You don't have anything to worry about," I said to the first boy. "The world's not going to end without you, okay?"

"It *will* end without him," his brother insisted. "He'll end without the world."

"Dad, do you know what time the world is going to die?"

"I have no idea when the world is going to die. It's a long time away."

"What's going to end the world, anyway?"

"Well, there are different theories about it," I said.

"What's one of them?"

Well, I said, the sun, which is constantly expanding, might one day grow so large that it engulfs the Earth. "But that is so far into the future that we can't even imagine it," I said.

"What's the second?"

"A black hole is going to suck us up," his brother said.

"Yes, maybe a black hole will suck us up," I said.

"The third?"

I explained how the universe began as a speck and then exploded and is now enormous, but that eventually it might stop expanding and perhaps even shrink and turn back into a speck. "So we'd get crushed into the speck," I said.

"Really?"

"Maybe," I said.

"Is that going to happen in a long, long time?"

"Yes, that's a long time away."

"So we won't be living then."

"No, we won't be," I said.

"Dad, what's another theory?"

"Let's think of one more, then let's think about bedtime," I said.

"Daddy, once the world is a speck, is it possible that it might explode again?"

"Sure, that's possible. It might start all over."

"Probably not," he said.

"Maybe not," I said. "It's interesting to think about, though."

Lately the boys' gravest concern involves my parents. My mother is in her late eighties and my father is past ninety, and they live several hours away in the house where I grew up. They are marvels of human biology, more so with each passing day. They garden, they sing in their church choir, they work out together every week with a trainer at the gym. They have activities: reading group, camera club, crosswords, movies. They still drive, which troubles me. We try to visit them often with the boys, but it isn't often enough.

A couple of summers ago I went with my father and the boys to the state fair. It's an excursion I've made with my parents nearly every year since I was small. The fair is held for several days at the end of August and into September, on a vast fairground of pavilions and booths. There are rooster-crowing contests and udder-judging contests, flower shows, quilting displays, a butterfly exhibit, row after row of heirloom rabbits and heirloom pigeons, a bearded guy with a woodworking spiel, vendors of blenders and maple-flavored cotton candy. There's a midway with nausea-inducing rides and dubious games of skill. Always there's the butter sculpture.

We took a shuttle bus from Shoppingtown Mall to avoid the hassle of parking. My father started talking about the war. He was drafted in 1944 but his eyesight was poor, so he was passed over for combat, a fact to which my siblings and I may well owe everything. Instead, for several months just after the war ended, he was stationed outside Paris in a military hospital and worked as a clerk. On the weekends he and his pals would go into the city, where they sold their ration of cigarettes and bought perfume and stockings to sell to the guys back on base. All the while, he said, he was learning French, turning it over in his mind. Sometimes he would step onto a bus or be walking somewhere and a French phrase would suddenly pop into his head, as if he were rehearsing for a play.

Lately, he said, he has a new inner monologue, about how old he is and about the friends who are leaving. Dying, he means; my parents have lost several close friends in the past few years. He mentioned the prescription eyedrops he takes. Sometimes, he said, he'll pick up the bottle and think about the miracle that is an eye and that both of his manage to still work. Sometimes he has these thoughts while he's sitting on the toilet, he said, and that's interesting too—how it all goes in and comes out, passing through and adding to the living machinery that is us, until it doesn't anymore.

He has a recurring dream in which he's a boy in the front seat of a car that his father is driving. In one version they're descending from the mountains toward a plain and he can see that at the bottom the road they're on branches into several roads, spreading out and going in all directions, and he's beginning to worry about which one is the right one and where he might be headed.

For some weeks, the man at the watch repair shop had been leaving phone messages: my wristwatch was fixed, when could I come pick it up? His latest message said that if I didn't come for my watch soon he would sell it. So one autumn day, several months after I'd dropped off the watch, I took the train into Grand Central and walked over to Fifth Avenue to reclaim it.

When I got to the shop, the repairman was at a desk peering at a watch through a jeweler's glass. He looked up and recognized me, then found a small plastic bag containing my wristwatch and handed it to me. Nobody was waiting, so I asked if he had maybe fifteen minutes to tell me how he had become a watch repairman. "Fifteen minutes?" he said, in a heavy accent. "Why you need fifteen minutes? I can tell you everything in five minutes."

He had grown up in Ukraine. At fifteen, he told his parents that he didn't want to go to school anymore; he wanted to do something but he didn't know what. Someone suggested watchmaking, so that's what he did. In those days, in postwar Russia, watch parts were hard to come by, so his job often involved crafting them by hand. Nowadays, he said, watch manufacturers rely on parts that are specific to their brands, but occasionally a repair calls for a part that is just as easy for him to make himself. He stepped over to his desk and came back with a Rolex with its back open, revealing a microcosm of spinning gears. He pointed with pride to a tiny post that held the watch's balance wheel in place; he had made it by hand. I asked what aspect of the job most satisfied him. He gave me a puzzled look. "Fixing watches," he said. "Somebody brings it in, it doesn't work, I fix it, then it works—that's satisfying."

I paid the bill and made my way back to Grand Central. I had time before my train, so I sat at a café table and took out my watch. The repairman told me that he'd made it waterproof; I noticed that it was two minutes ahead of the time on my phone. I slipped it onto my wrist and felt its old weight again, then promptly forgot about it.

I looked around. Two older women sat on stools at a soda counter, chatting. Nearby, at a table, a French couple and their two children were eating ice cream cones. A priest hurried past. I saw a woman writing in her notebook and a man, alone, with his elbow on the table and his hand under his chin, asleep. All around people were looking at their phones, or talking on their phones, or talking with each other, and everywhere a suffusing drone of business and conversation—the sound of an intensely social species working to connect and to synchronize within itself.

The effect was soothing; I had been working from home for the past few months and it had been a while since I'd felt like a cog in anything. I noticed my watch: twelve minutes until my train. Susan and I had been taking turns running the dinnertime ritual and putting the boys to bed; tonight was my night. I used to dread it, because the boys resisted it; the arc from bath to tooth brushing to pajamas to story time should be a simple narrative but they had created an epic, some hybrid of Homer and Vonnegut, digressive and anxious. By the time it finally ended, with the lights out and two boys asleep, I often was asleep too, on their floor.

One theory in the child-rearing books is that young children resist sleep because they fear it; they're still new to the experience of waking up the next morning, so good night feels too much like good-bye. But in recent weeks something had shifted; the boys had begun to embrace sleep, and our evening wind-down was less fraught, more pleasant. For a while, one boy needed to have his back stroked before he could relax enough to drift off. Now the activity mostly served to reassure me. He'd tolerate it for a minute or two and then whisper diplomatically, "You can go away now."

. . .

I knew it was time to finish the book when my boys were old enough to start asking about it. "What's it about? Why is it taking so long?" They had opinions about how many pages I should be writing each day and about how many words should be on each page, and at dinner they would ask me if I had met my word count and offer perspective such as, "J. K. Rowling writes a lot faster than *that*." One day, from the back seat of the car, they proposed book titles. One suggested *Time Is Confusing*, which struck me as apt but maybe not so inviting. The other suggested *The People That Time Forgot*, which sounded like a terrific adventure story but also, perhaps, like an unwitting reference to himself and the other neglected members of his family.

Years ago, long before I had children or was even married, a friend with children said, "The thing about having kids is, after a while you forget what it was like before you had them." The idea was inconceivable. I couldn't envisage a future self whose comings and goings were completely circumscribed, apparently happily, by the wants and needs of someone half my size. But that's what happens. As I grew into the role, I sometimes felt as if I were taking apart a ship and using the wood to build a ship for someone else. Plank by plank I dismantled and rearranged myself until only one thing remained from my life before children: the book. There was less time than ever for it, so it took what little time there was—evenings, weekends, summers, holidays. It was all-consuming in a haphazard way, which seemed normal because it had been that way before too. It couldn't continue. And yet, to enter it on a rainy Saturday or in the late hours of the night was like entering a warm crawl space in the attic. It was tempting to imagine that the project might never end. One could argue that the book, having taken so long, had become another child I didn't want to let go of, one whose fate I could actually control.

I also wondered if my strategy for reckoning with time had been too clever by half. To Augustine, a syllable, sentence, or stanza in motion was the embodiment of time; unfurling, it stretches between past and future, memory and expectation, spanning now and its container, the self. "What is true of the poem as a whole is true equally of its individual stanzas and syllables," he wrote. "The same is true of the whole long performance, in which this poem may be a single item." Hypothetically, the same is true of a book: as long as it remained in motion, the author's present would never end. You can see where this logic is headed. Immortality was a book that was perpetually unfinished.

So much—all that matters, Augustine wrote—unfolds in a sentence. Somewhere along the way I'd lost track of the present tense and the thread of the message: that the soul (I might as well call it that at this point) is in its speaking, in the sentence neither completed nor unsaid but that is even now still falling from one's lips.

It isn't summer, or the end of it, until I've been to the beach. I don't mean a beach by a lake, where the waves loll and it's mucky underfoot and you can see weeds growing up from the bottom. I need an honest ocean beach, with white sand dunes and a sea breeze that snaps the lifeguard's flag, where your hair gets salty just from your sitting there and the surf slams and tosses foam and reminds you that there's nothing but sea between you and Normandy.

For a long time our boys were fascinated and frightened by this sort of beach, as one should be, but I knew that summer had arrived in an eternal sense the year they came to love it. They were five. It was Labor Day weekend, that radiant diapause between languor and regimentation, when the days shed their names and hint at something everlasting. A hurricane had come and gone, leaving sun and froth. The boys spent the early part of an afternoon learning how to be properly dashed by waves in that way that leaves seawater streaming from your nose. Then the tide started going out and it was time to build sand castles.

Here is an essential human pleasure: to fill your hand with sand, turn it over, and call it architecture. We picked a spot at the lowest sustainable point below the tide line. This was prime floodplain real estate—flat, with perfectly moist sand, but also exposed; our work would be the first to fall when the tide returned. In only a few minutes one boy had flung together a hamlet of sand mounds protected by a low, curving wall. I dug a moat in front of it, to slow the first waves, whenever they came, and built a breakwater in front of that. He looked on with gleeful amazement. "We've never had this much time!" he exclaimed. He meant, I think, that he had never been so close to big waves—the tide was still receding—and yet felt so unthreatened and unrushed. I noticed younger parents higher up on the beach. "Look at our little town," the boy said, proudly. He said again, "We've never had this much time!"

Nietzsche argued—actually, the psychoanalyst Stephen Mitchell argued that Nietzsche argued—that one can gauge a man's relationship to time by the way he builds a sand castle. The first man will proceed hesitantly, intent on craft but meanwhile fretting about the inevitable return of the waves and shocked by his loss when they finally arrive. A second man won't even start building: why bother if the tide will only destroy it? The third—the paragon of manhood, in Nietzsche's view—embraces the unavoidable and throws himself into the work regardless, joyful but not oblivious.

I would like to think that I belong to this third category, but I'm lucky if I'm in the first. I noticed that my other son, against my gentle advice, had started his construction project—a small, sculpted mound—in front of the breakwater and the protective wall of the sand town. The first errant wave reduced it to a wet lump, and him to tears. He started a second hermitage, which soon crumbled, then another. Nietzsche should have a fourth category for him, I thought—the man who is slightly apart yet fiercely attached. By now the tide had turned with vigor and the first low waves rushed up the beach. He was the first casualty; then the waves overran my breakwater and my moat and lashed the wall of the town; then they curled behind the wall and flooded the streets. The first boy stood behind the ramparts facing the tide, his arms outstretched, the grin of ages on his face.

"The end is here! The end is here!"

He was a giant. He had never looked so happy, and I envied him.

ACKNOWLEDGMENTS

This book was made possible with the support of the Solomon H. Guggenheim Foundation and the MacDowell Colony.

NOTE ON SOURCES

The literature on time is boundless. Since history began writers have poured forth their ideas about the subject, many of them thoughtful or provocative in an anecdotal sort of way but relatively few of them, until recently, scientific. At the risk—although in fact with the express aim—of ignoring great swaths of philosophy and religious thought, I've focused here mainly on the effort to plumb the human relationship to time through experiments, an endeavor that began in earnest about a century and a half ago. I do this knowing full well that even well-intentioned experiments can be poorly designed, or can yield vague or conflicting results, or may address such a narrow aspect of our temporal experience that it's hard to say whether they apply beyond the confines of the laboratory in which they were conceived.

Moreover, even this subset of the literature, loosely confined to experiments and their results, is voluminous. Early on I encountered the life's work of Julius T. Fraser, arguably the foremost authority on the interdisciplinary study of time. In 1966 Fraser founded the International Society for the Study of Time, which every three years convened a conference and brought together time researchers of all stripes, from physicists, Kantian philosophers, and medieval historians to neurobiologists, anthropologists, and Proust scholars. Fraser gradually collected the papers into a ten-volume series of eclectic yet probing books, *The Study of Time,* and he wrote or edited several others, including *Time, the Familiar Stranger* and *The Voices of Time: A Cooperative Survey of Man's View of Time as Expressed by the Humanities.* Frederick Turner, the poet and scholar, described Fraser admiringly as "a sort of combination of Einstein, Yoda, Gandalf, Dr. Johnson, Socrates, the Old Testament God, and Groucho Marx." I'd heard that Fraser had retired to Connecticut, but by the time I'd read enough of his work to feel confident approaching him, he had died, at the age of eighty-seven.

This book should not be confused with an encyclopedia of time. (There are at least two of those: one, published in 1994, is seven hundred pages long and weighs three pounds; the second, published in 2009, contains sixteen hundred pages in three volumes and weighs eleven pounds.) I can guarantee that these pages do not answer your every last question about time. Instead, with the interest of both the reader and the writer in mind, I limited myself to what seemed humanly possible: a brief survey of that portion of the field that is of greatest interest to me and, I hope, by extension, to you. For those wanting to read more, my key sources follow. Watch out for rabbit holes.

BIBLIOGRAPHY

FORWARD

Augustine. *The Confessions*. Translated by Maria Boulding. New York: Vintage Books, 1998.

Gilbreth, Frank B., and Lillian Moller Gilbreth. *Fatigue Study: the Elimination of Humanity's Greatest Unnecessary Waste, a First Step in Motion Study*. New York: Macmillan Company, 1919.

Gilbreth, Frank B., and Robert Thurston Kent. *Motion Study, a Method for Increasing the Efficiency of the Workman*. New York: D. Van Nostrand, 1911.

Gleick, James. *Faster: The Acceleration of Just about Everything*. New York: Pantheon Books, 1999.

James, William. "Does Consciousness Exist?" *Journal of Philosophy, Psychology and Scientific Methods* 1, no. 18 (1904).

Lakoff, George, and Mark Johnson. *Philosophy in the Flesh: The Embodied Mind and Its Challenge to Western Thought*. New York: Basic Books, 1999.

Robinson, John P., and Geoffrey Godbey. *Time for Life: The Surprising Ways Americans Use Their Time*. University Park, PA: Pennsylvania State University Press, 1997.

THE HOURS

Adam, Barbara. *Timewatch: The Social Analysis of Time*. Cambridge, UK: Polity Press, 1995.

Arias, Elisa Felicitas. "The Metrology of Time." *Philosophical Transactions. Series A, Mathematical, Physical, and Engineering Sciences* 363, no. 1834 (2005): 2289–2305.

Battersby, S. "The Lady Who Sold Time." *New Scientist,* February 25–March 3, 2006, 52–53.

Brann, Eva T. H. *What, Then, Is Time?* Lanham, MD: Rowman & Littlefield, 1999.

Cockell, Charles S., and Lynn J. Rothschild. "The Effects of Ultraviolet Radiation A and B on Diurnal Variation in Photosynthesis in Three Taxonomically and Ecologically Diverse Microbial Mats." *Photochemisty and Photobiology* 69 (1999): 203–10.

Friedman, William J. "Developmental and Cognitive Perspectives on Humans' Sense of the Times of Past and Future Events." *Learning and Motivation* 36, no. 2 (2005): 145–58.

Goff, Jacques Le. *Time, Work, and Culture in the Middle Ages.* Chicago: University of Chicago Press, 1980.

Koriat, Asher, and Baruch Fischhoff. "What Day Is Today? An Inquiry into the Process of Temporal Orientation." *Memory and Cognition* 2, no. 2 (1974): 201–5.

Parker, Thomas E., and Demetrios Matsakis. "Time and Frequency Dissemination: Advances in GPS Transfer Techniques." *GPS World,* November 2004, 32–38.

Rifkin, Jeremy. *Time Wars: The Primary Conflict in Human History.* New York: H. Holt, 1987.

Rooney, David. *Ruth Belville: The Greenwich Time Lady.* London: National Maritime Museum, 2008.

Zerubavel, Eviatar. *Hidden Rhythms: Schedules and Calendars in Social Life.* Chicago: University of Chicago Press, 1981.

———. *The Seven Day Circle: The History and Meaning of the Week.* New York: Free Press, 1985.

THE DAYS

Alden, Robert. "Explorer Tells of Cave Ordeal." *New York Times,* September 20, 1962.

Antle, Michael C., and Rae Silver. "Orchestrating Time: Arrangements of the Brain Circadian Clock." *Trends in Neurosciences* 28, no. 3 (2005): 145–51.

Basner, Mathias, David F. Dinges, Daniel Mollicone, Adrian Ecker, Christopher W. Jones, Eric C. Hyder, Adrian Di, et al. "Mars 520-D Mission

Simulation Reveals Protracted Crew Hypokinesis and Alterations of Sleep Duration and Timing." *Proceedings of the National Academy of Sciences of the United States of America* 110, no. 7 (2012): 2635–40.

Bertolucci, Cristiano, and Augusto Foà. "Extraocular Photoreception and Circadian Entrainment in Nonmammalian Vertebrates." *Chronobiology International* 21, no. 4–5 (2004): 501–19.

Bradshaw, W. E., and C. M. Holzapfel. "Genetic Shift in Photoperiodic Response Correlated with Global Warming." *Proceedings of the National Academy of Sciences of the United States of America* 98, no. 25 (2001): 14509–11.

Bray, M. S., and M. E. Young. "Circadian Rhythms in the Development of Obesity: Potential Role for the Circadian Clock within the Adipocyte." *Obesity Reviews* 8, no. 2 (2007): 169–81.

Byrd, Richard Evelyn. *Alone: The Classic Polar Adventure.* New York: Kodansha International, 1995.

Castillo, Marina R., Kelly J. Hochstetler, Ronald J. Tavernier, Dana M. Greene, Abel Bult-ito. "Entrainment of the Master Circadian Clock by Scheduled Feeding." *American Journal of Physiology. Regulatory, Integrative and Comparative Physiology* 287 (2004): 551–55.

Cockell, Charles S., and Lynn J. Rothschild. "Photosynthetic Rhythmicity in an Antarctic Microbial Mat and Some Considerations on Polar Circadian Rhythms." *Antarctic Journal* 32 (1997): 156–57.

Coppack, Timothy, and Francisco Pulido. "Photoperiodic Response and the Adaptability of Avian Life Cycles to Environmental Change." *Advances in Ecological Research* 35 (2004): 131–50.

Covington, Michael F., and Stacey L. Harmer. "The Circadian Clock Regulates Auxin Signaling and Responses in Arabidopsis." *PLoS Biology* 5, no. 8 (2007): 1773–84.

Czeisler, C. A., J. S. Allan, S. H. Strogatz, J. M. Ronda, R. Sanchez, C. D. Rios, W. O. Freitag, G. S. Richardson, and R. E. Kronauer. "Bright Light Resets the Human Circadian Pacemaker Independent of the Timing of the Sleep-Wake Cycle." *Science* 233, no. 4764 (1986): 667–71.

Czeisler, Charles A., Jeanne F. Duffy, Theresa L. Shanahan, Emery N. Brown, F. Jude, David W. Rimmer, Joseph M. Ronda, et al. "Stability, Precision, and near-24-Hour Period of the Human Circadian Pacemaker." *Science* 284, no. 5423 (1999): 2177–81.

Dijk, D. J., D. F. Neri, J. K. Wyatt, J. M. Ronda, E. Riel, A. Ritz-De Cecco, R. J. Hughes, et al. "Sleep, Performance, Circadian Rhythms, and Light-Dark Cycles during Two Space Shuttle Flights." *American Journal of Physiology. Regulatory, Integrative and Comparative Physiology* 281, no. 5 (2001): R1647–64.

Dunlap, Jay C. "Molecular Bases for Circadian Clocks (Review)." *Cell* 96, no. 2 (1999): 271–90.

Figueiro, Mariana G., and Mark S. Rea. "Evening Daylight May Cause Adolescents to Sleep Less in Spring Than in Winter." *Chronobiology International* 27, no. 6 (2010): 1242–58.

Foer, Joshua. "Caveman: An Interview with Michel Siffre." *Cabinet Magazine* no. 30, Summer 2008, http://www.cabinetmagazine.org/issues/30/foer.php.

Foster, Russell G. "Keeping an Eye on the Time." *Investigative Ophthalmology* 43, no. 5 (2002): 1286–98.

Froy, Oren. "The Relationship between Nutrition and Circadian Rhythms in Mammals." *Frontiers in Neuroendocrinology* 28, no. 2–3 (2007): 61–71.

Golden, Susan S. "Meshing the Gears of the Cyanobacterial Circadian Clock." *Proceedings of the National Academy of Sciences* 101, no. 38 (2004): 13697–98.

———. "Timekeeping in Bacteria: The Cyanobacterial Circadian Clock." *Current Opinion in Microbiology* 6, no. 6 (2003): 535–40.

Golden, Susan S., and Shannon R. Canales. "Cyanobacterial Circadian Clocks: Timing Is Everything." *Nature Reviews. Microbiology* 1, no. 3 (2003): 191–99.

Golombek, Diego A., Javier A. Calcagno, and Carlos M. Luquet. "Circadian Activity Rhythm of the Chinstrap Penguin of Isla Media Luna, South Shetland Islands, Argentine Antarctica." *Journal of Field Ornithology* 62, no. 3 (1991): 293–428.

Gooley, J. J., J. Lu, T. C. Chou, T. E. Scammell, and C. B. Saper. "Melanopsin in Cells of Origin of the Retinohypothalamic Tract." *Nature Neuroscience* 4, no. 12 (2001): 1165.

Gronfier, Claude, Kenneth P. Wright, Richard E. Kronauer, and Charles A. Czeisler. "Entrainment of the Human Circadian Pacemaker to Longer-than-24-H Days." *Proceedings of the National Academy of Sciences of the United States of America* 104, no. 21 (2007): 9081–86.

Hamermesh, Daniel S., Caitlin Knowles Myers, and Mark L. Pocock. "Cues for Timing and Coordination: Latitude, Letterman, and Longitude." *Journal of Labor Economics* 26, no. 2 (2008): 223–46.

Hao, H., and S. A. Rivkees. "The Biological Clock of Very Premature Primate Infants Is Responsive to Light." *Proceedings of the National Academy of Sciences of the United States of America* 96, no. 5 (1999): 2426–29.

Hellwegera, Ferdi L. "Resonating Circadian Clocks Enhance Fitness in Cyanobacteria in Silico." *Ecological Modelling* 221, no. 12 (2010): 1620–29.

Johnson, Carl Hirschie, and Martin Egli. "Visualizing a Biological Clockwork's Cogs." *Nature Structural and Molecular Biology* 11, no. 7 (2004): 584–85.

Johnson, Carl Hirschie, Tetsuya Mori, and Yao Xu. "A Cyanobacterial Circadian Clockwork." *Current Biology* 18, no. 17 (2008): R816–25.

Kohsaka, Akira, and Joseph Bass. "A Sense of Time: How Molecular Clocks Organize Metabolism." *Trends in Endocrinology and Metabolism* 18, no. 1 (2007): 4–11.

Kondo, T. "A Cyanobacterial Circadian Clock Based on the Kai Oscillator." In *Cold Spring Harbor Symposia on Quantitative Biology* 72, (2007): 47–55.

Konopka, R. J., and S. Benzer. "Clock Mutants of *Drosophila Melanogastermelanogaster*." *Proceedings of the National Academy of Sciences of the United States of America* 68, no. 9 (1971): 2112–16.

Lockley, Steven W., and Joshua J. Gooley. "Circadian Photoreception: Spotlight on the Brain." *Current Biology* 16, no. 18 (2006): R795–97.

Lu, Weiqun, Qing Jun Meng, Nicholas J. C. Tyler, Karl-Arne Stokkan, and Andrew S. I. Loudon. "A Circadian Clock Is Not Required in an Arctic Mammal." *Current Biology* 20, no. 6 (2010): 533–37.

Lubkin, Virginia, Pouneh Beizai, and Alfredo A. Sadun. "The Eye as Metronome of the Body." *Survey of Ophthalmology* 47, no. 1 (2002): 17–26.

Mann, N. P. "Effect of Night and Day on Preterm Infants in a Newborn Nursery: Randomised Trial." *British Medical Journal* 293 (November 1986): 1265–67.

McClung, Robertson. "Plant Circadian Rhythms." *Plant Cell* 18 (April 2006): 792–803.

Meier-Koll, Alfred, Ursula Hall, Ulrike Hellwig, Gertrud Kott, and Verena Meier-Koll. "A Biological Oscillator System and the Development of Sleep–Waking Behavior during Early Infancy." *Chronobiologia* 5, no. 4 (1978): 425–40.

Menaker, Michael. "Circadian Rhythms. Circadian Photoreception." *Science* 299, no. 5604 (2003): 213–14.

Mendoza, Jorge. "Circadian Clocks: Setting Time by Food." *Journal of Neuroendocrinology* 19, no. 2 (2007): 127–37.

Mills, J. N., D. S. Minors, J. M. Waterhouse, and M. Manchester. "The Circadian Rhythms of Human Subjects without Timepieces or Indication of the Alternation of Day and Night." *Journal of Physiology* 240, no. 3 (1974): 567–94.

Mirmiran, Majid, J. H. Kok, K. Boer, and H. Wolf. "Perinatal Development of Human Circadian Rhythms: Role of the Foetal Biological Clock." *Neuroscience and Biobehavioral Reviews* 16, no. 3 (1992): 371–78.

Mittag, Maria, Stefanie Kiaulehn, and Carl Hirschie Johnson. "The Circadian Clock in *Chlamydomonas Reinhardtiireinhardtii*: What Is It For? What Is It Similar To?" *Plant Physiology* 127, no. 2 (2005): 399–409.

Monk, T. H., K. S. Kennedy, L. R. Rose, and J. M. Linenger. "Decreased Human Circadian Pacemaker Influence after 100 Days in Space: A Case Study." *Psychosomatic Medicine* 63, no. 6 (2001): 881–85.

Monk, Timothy H., Daniel J. Buysse, Bart D. Billy, Kathy S. Kennedy, and Linda M. Willrich. "Sleep and Circadian Rhythms in Four Orbiting Astronauts." *Journal of Biological Rhythms* 13 (June 1998): 188–201.

Murayama, Yoriko, Atsushi Mukaiyama, Keiko Imai, Yasuhiro Onoue, Akina Tsunoda, Atsushi Nohara, Tatsuro Ishida, et al. "Tracking and Visualizing the Circadian Ticking of the Cyanobacterial Clock Protein KaiC in Solution." *EMBO Journal* 30, no. 1 (2011): 68–78.

Nikaido, S. S., and C. H. Johnson. "Daily and Circadian Variation in Survival from Ultraviolet Radiation in *Chlamydomonas Reinhardtiireinhardtii*." *Photochemistry and Photobiology* 71, no. 6 (2000): 758–65.

O'Neill, John S., and Akhilesh B. Reddy. "Circadian Clocks in Human Red Blood Cells." *Nature* 469, no. 7331 (2011): 498–503.

Ouyang, Yan, Carol R. Andersson, Takao Kondo, Susan S. Golden, and Carl Hirschie Johnson. "Resonating Circadian Clocks Enhance Fitness in Cyanobacteria" *Proceedings of the National Academy of Sciences of the United States of America* 95 (July 1998): 8660–64.

Palmer, John D. *The Living Clock: The Orchestrator of Biological Rhythms.* Oxford: Oxford University Press, 2002.

Panda, Satchidananda, John B. Hogenesch, and Steve A. Kay. "Circadian Rhythms from Flies to Human." *Nature* 417, no. 6886 (2002): 329–35.

Pöppel, Ernst. "Time Perception." In *Handbook of Sensory Physiology*. Vol. 8, *Perception*, edited by R. Held, H. W. Leibowitz, and H. L. Teubner. Berlin: Springer-Verlag, 1978, 713–29.

Ptitsyn, Andrey A., Sanjin Zvonic, Steven A. Conrad, L. Keith Scott, Randall L. Mynatt, and Jeffrey M Gimble. "Circadian Clocks Are Resounding in Peripheral Tissues." *PLoS Computational Biology* 2, no. 3 (2006): 126–35.

Ptitsyn, Andrey A., Sanjin Zvonic, and Jeffrey M. Gimble. "Digital Signal Processing Reveals Circadian Baseline Oscillation in Majority of Mammalian Genes." *PLoS Computational Biology* 3, no. 6 (2007): 1108–14.

Ramsey, Kathryn Moynihan, Biliana Marcheva, Akira Kohsaka, and Joseph Bass. "The Clockwork of Metabolism." *Annual Review of Nutrition* 27, (2007): 219–40.

Reppert, S. M. "Maternal Entrainment of the Developing Circadian System." *Annals of the New York Academy of Sciences* 453 (1985): 162–69, fig. 2.

Revel, Florent G., Annika Herwig, Marie-Laure Garidou, Hugues Dardente, Jérôme S. Menet, Mireille Masson-Pévet, Valérie Simonneaux, Michel Saboureau, and Paul Pévet. "The Circadian Clock Stops Ticking during Deep Hibernation in the European Hamster." *Proceedings of the National Academy of Sciences of the United States of America* 104, no. 34 (2007): 13816–20.

Rivkees, Scott A. "Developing Circadian Rhythmicity in Infants." *Pediatrics* 112, no. 2 (2003): 373–81

Rivkees, Scott A., P. L. Hofman, and J. Fortman. "Newborn Primate Infants Are Entrained by Low Intensity Lighting." *Proceedings of the National Academy of Sciences of the United States of America* 94, no. 1 (1997): 292–97.

Rivkees, Scott A., Linda Mayes, Harris Jacobs, and Ian Gross. "Rest-Activity Patterns of Premature Infants Are Regulated by Cycled Lighting." *Pediatrics* 113, no. 4 (2004): 833–39.

Rivkees, Scott A., and S. M. Reppert. "Perinatal Development of Day-Night Rhythms in Humans." *Hormone Research* 37, Supplement 3 (1992): 99–104.

Roenneberg, Till, Karla V. Allebrandt, Martha Merrow, and Céline Vetter. "Social Jetlag and Obesity." *Current Biology* 22, no. 10 (2012): 939–43.

Roenneberg, Till, and Martha Merrow. "Light Reception: Discovering the Clock-Eye in Mammals." *Current Biology* 12, no. 5 (2002): R163–65.

Rubin, Elad B., Yair Shemesh, Mira Cohen, Sharona Elgavish, Hugh M. Robertson, and Guy Bloch. "Molecular and Phylogenetic Analyses Reveal Mammalian-like Clockwork in the Honey Bee (*Apis Melliferamellifera*) and Shed New Light on the Molecular Evolution of the Circadian Clock." *Genome Research* 16, no. 11 (2006): 1352–65.

Scheer, Frank A. J. L., Michael F. Hilton, Christos S. Mantzoros, and Steven A. Shea. "Adverse Metabolic and Cardiovascular Consequences of Circadian Misalignment." *Proceedings of the National Academy of Sciences of the United States of America* 106, no. 11 (2009): 4453–58.

Scheer, Frank A. J. L., Kenneth P. Wright, Richard E. Kronauer, and Charles A. Czeisler. "Plasticity of the Intrinsic Period of the Human Circadian Timing System." *PLoS ONE* 2, no. 8 (2007): e721.

Siffre, Michel. *Hors du temps: L'expérience du 16 juillet 1962 au fond du gouffre de Scarasson par celui qui l'a vécue.* Paris: R. Julliard, 1963.

———. "Six Months Alone in a Cave." *National Geographic*, March 1975, 426–35.

Skuladottir, Arna, Marga Thome, and Alfons Ramel. "Improving Day and Night Sleep Problems in Infants by Changing Day Time Sleep Rhythm: A Single Group before and after Study." *International Journal of Nursing Studies* 42, no. 8 (2005): 843–50.

Sorek, Michal, Yosef Z. Yacobi, Modi Roopin, Ilana Berman-Frank, and Oren Levy. "Photosynthetic Circadian Rhythmicity Patterns of Symbiodinium, the Coral Endosymbiotic Algae." *Proceedings. Biological Sciences / The Royal Society* 280 (2013): 20122942.

Stevens, Richard G., and Yong Zhu. "Electric Light, Particularly at Night, Disrupts Human Circadian Rhythmicity: Is That a Problem?" *Philosophical Transactions of the Royal Society of London. Series B, Biological Sciences* 370, no. 1667 (March 16, 2015): 20140120.

Stokkan, Karl-Arne, Shin Yamazaki, Hajime Tei, Yoshiyuki Sakaki, and Michael Menaker. "Entrainment of the Circadian Clock in the Liver by Feeding." *Science* 291 (2001): 490–93.

Strogatz, Steven H. *Sync: The Emerging Science of Spontaneous Order.* New York: Hyperion, 2003.

Suzuki, Lena, and Carl Hirschie Johnson. "Algae Know the Time of Day: Circadian and Photoperiodic Programs." *Journal of Phycology* 37, no. 6 (2001): 933–42.

Takahashi, Joseph S., Kazuhiro Shimomura, and Vivek Kumar. "Searching for Genes Underlying Circadian Rhythms." *Science* 322 (November 7, 2008): 909–12.

Tavernier, Ronald J., Angela L. Largen, and Abel Bult-ito. "Circadian Organization of a Subarctic Rodent, the Northern Red-Backed Vole (*Clethrionomys Rutilusrutilus*)." *Journal of Biological Rhythms* 19, no. 3 (2004): 238–47.

United States Congress, Office of Technology Assessment. *Biological Rhythms: Implications for the Worker.* Washington, D.C.: U.S. Government Printing Office, 1991.

Van Oort, Bob E. H., Nicholas J. C. Tyler, Menno P. Gerkema, Lars Folkow, Arnoldus Schytte Blix, and Karl-Arne Stokkan. "Circadian Organization in Reindeer." *Nature* 438, no. 7071 (2005): 1095–96.

Weiner, Jonathan. *Time, Love, Memory: A Great Biologist and His Quest for the Origins of Behavior.* New York: Knopf, 1999.

Wittmann, Marc, Jenny Dinich, Martha Merrow, and Till Roenneberg. "Social Jetlag: Misalignment of Biological and Social Time." *Chronobiology International* 23, no. 1–2 (2006): 497–509.

Woelfle, Mark A., Yan Ouyang, Kittiporn Phanvijhitsiri, and Carl Hirschie Johnson. "The Adaptive Value of Circadian Clocks: An Experimental Assessment in Cyanobacteria." *Current Biology* 14 (August 24, 2004): 1481–86.

Wright, Kenneth P., Andrew W. McHill, Brian R. Birks, Brandon R. Griffin, Thomas Rusterholz, and Evan D. Chinoy. "Entrainment of the Human Circadian Clock to the Natural Light-Dark Cycle." *Current Biology* 23, no. 16 (2013): 1554–58.

Xu, Yao, Tetsuya Mori, and Carl Hirschie Johnson. "Cyanobacterial Circadian Clockwork: Roles of KaiA, KaiB and the KaiBC Promoter in Regulating KaiC." *EMBO Journal* 22, no. 9 (2003): 2117–26.

Zivkovic, Bora, "Circadian Clock without DNA: History and the Power of Metaphor." *Observations* (blog), *Scientific American* (2011): 1–25.

THE PRESENT

Allport, D. A. "Phenomenal Simultaneity and the Perceptual Moment Hypothesis." *British Journal of Psychology* 59, no. 4 (1968): 395–406.

Baugh, Frank G., and Ludy T. Benjamin. "Walter Miles, Pop Warner, B. C. Graves, and the Psychology of Football." *Journal of the History of the Behavioral Sciences* 42, Winter (2006): 3–18.

Blatter, Jeremy. "Screening the Psychological Laboratory: Hugo Münsterberg, Psychotechnics, and the Cinema, 1892–1916." *Science in Context* 28, no. 1 (2015): 53–76.

Boring, Edwin Garrigues. *A History of Experimental Psychology.* New York: Appleton-Century-Crofts, 1950.

———. *Sensation and Perception in the History of Experimental Psychology.* New York: Appleton-Century-Crofts, 1942.

Buonomano, Dean V., Jennifer Bramen, and Mahsa Khodadadifar. "Influence of the Interstimulus Interval on Temporal Processing and Learning: Testing the State-Dependent Network Model." *Philosophical Transactions of the Royal Society of London. Series B, Biological Sciences* 364, no. 1525 (2009): 1865–73.

Cai, Mingbo, David M. Eagleman, and Wei Ji Ma. "Perceived Duration Is Reduced by Repetition but Not by High- Level Expectation." *Journal of Vision* 15, no. 13 (2015): 1–17.

Cai, Mingbo, Chess Stetson, and David M. Eagleman. "A Neural Model for Temporal Order Judgments and Their Active Recalibration: A Common Mechanism for Space and Time?" *Frontiers in Psychology* 3 (November 2012): 470.

Campbell, Leah A., and Richard A. Bryant. "How Time Flies: A Study of Novice Skydivers." *Behaviour Research and Therapy* 45, no. 6 (2007): 1389–92.

Canales, Jimena. "Exit the Frog, Enter the Human: Physiology and Experimental Psychology in Nineteenth-Century Astronomy." *British Journal for the History of Science* 34, no. 2 (2001): 173–97.

———. *A Tenth of a Second: A History.* Chicago: University of Chicago Press, 2009.

Dierig, Sven. "Engines for Experiment: Labor Revolution and Industrial in the Nineteenth-Century City." In *Osiris.* Vol. 18, *Science and the City,* edited by Sven Dierig, Jens Lachmund, and Andrew Mendelsohn. University of Chicago Press, 2003, 116–34.

Dollar, John, director and producer. "Prisoner of Consciousness." *Equinox,* season 1, episode 3. Channel 4 (UK), aired August 4, 1986.

Duncombe, Raynor L. "Personal Equation in Astronomy." *Popular Astronomy* 53 (1945): 2–13, 63–76, 110–121.

Eagleman, David M. "How Does the Timing of Neural Signals Map onto the Timing of Perception?" In *Space and Time in Perception and Action,* edited by R. Nijhawan and B. Khurana. Cambridge, UK: Cambridge University Press, 2010, 216–31.

———. "Human Time Perception and Its Illusions." *Current Opinion in Neurobiology* 18, no. 2 (2008): 131–36.

———. "Motion Integration and Postdiction in Visual Awareness." *Science* 287, no. 5460 (2000): 2036–38.

———. "The Where and When of Intention." *Science* 303, no. 5661 (2004): 1144–46.

Eagleman, David M., and Alex O. Holcombe. "Causality and the Perception of Time." *Trends in Cognitive Sciences* 6, no. 8 (2002): 323–25.

Eagleman, David M., and Vani Pariyadath. "Is Subjective Duration a Signature of Coding Efficiency?" *Philosophical Transactions of the Royal Society of London. Series B, Biological Sciences* 364, no. 1525 (2009): 1841–51.

Eagleman, David M., P. U. Tse, Dean V. Buonomano, P. Janssen, A. C. Nobre, and A. O. Holcombe. "Time and the Brain: How Subjective Time Relates to Neural Time." *Journal of Neuroscience* 25, no. 45 (2005): 10369–71.

Efron, R. "The Duration of the Present." *Annals of the New York Academy of Sciences* 138 (February 1967): 712–29.

Ekirch, A. Roger. *At Day's Close: Night in Times Past.* New York: W. W. Norton, 2006.

Engel, Andreas K., Pascal Fries, P. König, Michael Brecht, and Wolf Singer. "Temporal Binding, Binocular Rivalry, and Consciousness." *Consciousness and Cognition* 8, no. 2 (1999): 128–51.

Engel, Andreas K., Pieter R. Roelfsema, Pascal Fries, Michael Brecht, and Wolf Singer. "Role of the Temporal Domain for Response Selection and Perceptual Binding." *Cerebral Cortex* 7, no. 6 (1997): 571–82.

Engel, Andreas K., and Wolf Singer. "Temporal Binding and the Neural Correlates of Sensory Awareness." *Trends in Cognitive Sciences* 5, no. 1 (2001): 16–25.

Friedman, William J. *About Time: Inventing the Fourth Dimension.* Cambridge, MA: MIT Press, 1990.

———. "Developmental and Cognitive Perspectives on Humans' Sense of the Times of Past and Future Events." *Learning and Motivation* 36, no. 2 Special Issue (2005): 145–58.

———. "Developmental Perspectives on the Psychology of Time." In *Psychology of Time,* edited by Simon Grondin. Bingley, UK: Emerald, 2008, 345–66.

———. "The Development of Children's Knowledge of Temporal Structure." *Child Development* 57, no. 6 (1986): 1386–1400.

———. "The Development of Children's Knowledge of the Times of Future Events." *Child Development* 71, no. 4 (2000): 913–32.

———. "The Development of Children's Understanding of Cyclic Aspects of Time." *Child Development* 48, no. 4 (1977): 1593–99.

———. "The Development of Infants' Perception of Arrows of Time." *Infant Behavior and Development* 19, Supplement 1 (1996): 161.

Friedman, William J., and Susan L. Brudos. "On Routes and Routines: The Early Development of Spatial and Temporal Representations." *Cognitive Development* 3, no. 2 (1988): 167–82.

Galison, Peter L. *Einstein's Clocks and Poincaré's Maps: Empires of Time.* New York: W. W. Norton, 2003.

Galison, Peter L., and D. Graham Burnett. "Einstein, Poincaré and Modernity: A Conversation." *Time* 132, no. 2 (2009): 41–55.

Gillings, Annabel, director and producer. "Daytime." *Time,* episode 1. BBC Four, aired on July 30, 2007.

Granier-Deferre, Carolyn, Sophie Bassereau, Aurélie Ribeiro, Anne-Yvonne Jacquet, and Anthony J. Decasper. "A Melodic Contour Repeatedly Experienced by Human Near-Term Fetuses Elicits a Profound Cardiac Reaction One Month after Birth." *PloS One* 6, no. 2 (2011): e17304.

Green, Christopher D., and Ludy T. Benjamin. *Psychology Gets in the Game: Sport, Mind, and Behavior, 1880–1960.* Lincoln: University of Nebraska Press, 2009.

Haggard, P., S. Clark, and J. Kalogeras. "Voluntary Action and Conscious Awareness." *Nature Neuroscience* 5, no. 4 (2002): 382–85.

Hale, Matthew. *Human Science and Social Order: Hugo Münsterberg and the Origins of Applied Psychology.* Philadelphia: Temple University Press, 1980.

Helfrich, Hede. *Time and Mind II: Information Processing Perspectives.* Toronto: Hogrefe & Huber, 2003.

Hoerl, Christoph, and Teresa McCormack. *Time and Memory: Issues in Philosophy and Psychology.* Oxford: Clarendon Press, 2001.

James, William. *The Principles of Psychology.* London: Macmillan, 1901.

Jenkins, Adrianna C., C. Neil Macrae, and Jason P. Mitchell. "Repetition Suppression of Ventromedial Prefrontal Activity during Judgments of Self and Others." *Proceedings of the National Academy of Sciences of the United States of America* 105, no. 11 (2008): 4507–12.

Karmarkar, Uma R., and Dean V. Buonomano. "Timing in the Absence of Clocks: Encoding Time in Neural Network States." *Neuron* 53, no. 3 (2007): 427–38.

Kline, Keith A., and David M. Eagleman. "Evidence against the Temporal Subsampling Account of Illusory Motion Reversal." *Journal of Vision* 8, no. 4 (2008): 13.1–13.5.

Kline, Keith A., Alex O. Holcombe, and David M. Eagleman. "Illusory Motion Reversal Is Caused by Rivalry, Not by Perceptual Snapshots of the Visual Field." *Vision Research* 44, no. 23 (2004): 2653–58.

Kornspan, Alan S. "Contributions to Sport Psychology: Walter R. Miles and the Early Studies on the Motor Skills of Athletes." *Comprehensive Psychology* 3, no. 1, article 17 (2014): 1–11.

Kreimeier, Klaus, and Annemone Ligensa. *Film 1900: Technology, Perception, Culture.* New Burnet, UK: John Libbey, 2009.

Lejeune, Helga, and John H. Wearden. "Vierordt's 'The Experimental Study of the Time Sense' (1868) and Its Legacy." *European Journal of Cognitive Psychology* 21, no. 6 (2009): 941–60.

Levin, Harry, and Ann Buckler-Addis. *The Eye–Voice Span.* Cambridge, MA: MIT Press, 1979.

Lewkowicz, David J. "The Development of Intersensory Temporal Perception: An Epigenetic Systems/Limitations View." *Psychological Bulletin* 126, no. 2 (2000): 281–308.

———. "Development of Multisensory Temporal Perception." In *The Neural Bases of Multisensory Processes,* edited by M. M. Murray and M. T. Wallace. Boca Raton, FL: CRC Press/Taylor & Francis, 2012, 325–44.

———. "The Role of Temporal Factors in Infant Behavior and Development." In *Time and Human Cognition,* edited by I. Levin and D. Zakay. North-Holland: Elsevier Science Publishers, 1989, 1–43.

Lewkowicz, David J., Irene Leo, and Francesca Simion. "Intersensory Perception at Birth: Newborns Match Nonhuman Primate Faces and Voices." *Infancy* 15, no. 1 (2010): 46–60.

Leyden, W. von. "History and the Concept of Relative Time." *History and Theory* 2, no. 3 (1963): 263–85.

Lickliter, R., and L. E. Bahrick. "The Development of Infant Intersensory Perception: Advantages of a Comparative Convergent-Operations Approach." *Psychological Bulletin* 126, no. 2 (2000): 260–80.

Matthews, William J., and Warren H. Meck. "Time Perception: The Bad News and the Good." *Wiley Interdisciplinary Reviews: Cognitive Science* 5, no. 4 (2014): 429–46.

Matthews, William J., Devin B. Terhune, Hedderik Van Rijn, David M. Eagleman, Marc A. Sommer, and Warren H. Meck. "Subjective Duration as a Signature of Coding Efficiency: Emerging Links among Stimulus Repetition, Predictive Coding, and Cortical GABA Levels." *Timing & Time Perception Reviews* 1, no. 5 (2014): 1–5.

Münsterberg, Hugo, and Allan Langdale. *Hugo Münsterberg on Film: The Photoplay: A Psychological Study, and Other Writings.* New York: Routledge, 2002.

Myers, Gerald E. "William James on Time Perception." *Philosophy of Science* 38, no. 3 (1971): 353–60.

Neil, Patricia A., Christine Chee-Ruiter, Christian Scheier, David J. Lewkowicz, and Shinsuke Shimojo. "Development of Multisensory Spatial Integration and Perception in Humans." *Developmental Science* 9, no. 5 (2006): 454–64.

Nelson, Katherine. "Emergence of the Storied Mind." In *Language in Cognitive Development: The Emergence of the Mediated Mind.* Cambridge, UK: Cambridge University Press, 1996, 183–291.

———. "Emergence of Autobiographical Memory at Age 4." *Human Development* 35, no. 3 (1992): 172–77.

Nichols, Herbert. *The Psychology of Time.* New York: Henry Holt, 1891.

Nijhawan, Romi. "Visual Prediction: Psychophysics and Neurophysiology of Compensation for Time Delays." *Behavioral and Brain Sciences* 31, no. 2 (2008): 179–98; discussion 198–239.

Nijhawan, Romi, and Beena Khurana. *Space and Time in Perception and Action.* Cambridge, UK: Cambridge University Press, 2010.

Pariyadath, Vani, and David M. Eagleman. "Brief Subjective Durations Contract with Repetition." *Journal of Vision* 8, no. 16 (2008): 1–6.

———. "The Effect of Predictability on Subjective Duration." *PloS One* 2, no. 11 (2007): e1264.

Pariyadath, Vani, Mark H. Plitt, Sara J. Churchill, and David M. Eagleman. "Why Overlearned Sequences Are Special: Distinct Neural Networks for Ordinal Sequences." *Frontiers in Human Neuroscience* 6 (December 2012): 1–9.

Piaget, Jean. "Time Perception in Children." In *The Voices of Time: A Cooperative Survey of Man's Views of Time as Expressed by the Sciences and by the Humanities,* edited by Julius Thomas Fraser, Amherst, MA: University of Massachusetts Press, 1981, 202–16.

Plato. *Parmenides.* Translated by R. E. Allen. New Haven, CT: Yale University Press, 1998.

Pöppel, Ernst. "Lost in Time: A Historical Frame, Elementary Processing Units and the 3-Second Window." *Acta Neurobiologiae Experimentalis* 64, no. 3 (2004): 295–301.

———. *Mindworks: Time and Conscious Experience.* Boston: Harcourt Brace Jovanovich, 1988.

Purves, D., J. A. Paydarfar, and T. J. Andrews. "The Wagon Wheel Illusion in Movies and Reality." *Proceedings of the National Academy of Sciences of the United States of America* 93, no. 8 (1996): 3693–97.

Richardson, Robert D. *William James: In the Maelstrom of American Modernism: A Biography.* Boston: Houghton Mifflin, 2006.

Sacks, Oliver. "A Neurologist's Notebook: The Abyss." *The New Yorker,* September 24, 2007, 100–11.

Schaffer, Simon. "Astronomers Mark Time: Discipline and the Personal Equation." *Science in Context* 2, no. 1 (1988): 115–45.

Schmidgen, Henning. "Mind, the Gap: The Discovery of Physiological Time." In *Film 1900: Technology, Perception, Culture,* edited by K. Kreimeier and A. Ligensa, 53–65. New Burnet, UK: John Libbey, 2009.

———. "Of Frogs and Men: The Origins of Psychophysiological Time Experiments, 1850–1865." *Endeavour* 26, no. 4 (2002): 142–48.

———. "Time and Noise: The Stable Surroundings of Reaction Experiments, 1860–1890." *Studies in History and Philosophy of Biological and Biomedical Sciences* 34, no. 2 (2003): 237–75.

Scripture, Edward Wheeler. *Thinking Feeling Doing.* Meadville, PA: Flood and Vincent, 1895.

Solnit, Rebecca. *River of Shadows: Eadweard Muybridge and the Technological Wild West.* New York: Viking, 2003.

VanRullen, Rufin, and Christof Koch. "Is Perception Discrete or Continuous?" *Trends in Cognitive Sciences* 7, no. 5 (2003): 207–13.

Vatakis, Argiro, and Charles Spence. "Evaluating the Influence of the 'Unity Assumption' on the Temporal Perception of Realistic Audiovisual Stimuli." *Acta Psychologica* 127, no. 1 (2008): 12–23.

Wearing, Deborah. *Forever Today: A Memoir of Love and Amnesia.* London: Doubleday, 2005.

———. "The Man Who Keeps Falling in Love with His Wife." *The Telegraph*, January 12, 2005, http://www.telegraph.co.uk/news/health/3313452/The-man-who-keeps-falling-in-love-with-his-wife.html.

Wojtach, William T., Kyongje Sung, Sandra Truong, and Dale Purves. "An Empirical Explanation of the Flash-Lag Effect." *Proceedings of the National Academy of Sciences of the United States of America* 105, no. 42 (2008): 16338–43.

Wundt, Wilhelm. *An Introduction to Psychology.* Translated by Rudolf Pinter. London, 1912.

WHY TIME FLIES

Alexander, Iona, Alan Cowey, and Vincent Walsh. "The Right Parietal Cortex and Time Perception: Back to Critchley and the Zeitraffer Phenomenon," *Cognitive Neuropsychology* 22, no. 3 (May 2005): 306–15.

Allan, Lorraine, Peter D. Balsam, Russell Church, and Herbert Terrace. "John Gibbon (1934–2001) Obituary." *American Psychologist* 57, no. 6-7 (2002): 436–37.

Allman, Melissa J., and Warren H. Meck. "Pathophysiological Distortions in Time Perception and Timed Performance," *Brain* 135, no. 3 (2012): 656–77.

Allman, Melissa J., Sundeep Teki, Timothy D. Griffiths, and Warren H. Meck. "Properties of the Internal Clock: First- and Second-Order Principles of Subjective Time," *Annual Review of Psychology* 65 (2014): 743–71.

Angrilli, Alessandro, Paolo Cherubini, Antonella Pavese, and Sara Manfredini. "The Influence of Affective Factors on Time Perception." *Perception & Psychophysics* 59, no. 6 (1997): 972–82.

Arantes, Joana, Mark E. Berg, and John H. Wearden. "Females' Duration Estimates of Briefly-Viewed Male, but Not Female, Photographs Depend on Attractiveness." *Evolutionary Psychology* 11, no. 1 (2013): 104–19.

Arstila, Valtteri. *Subjective Time: The Philosophy, Psychology, and Neuroscience of Temporality.* Cambridge, MA: MIT Press, 2014.

Baer, Karl Ernst von: *"Welche Auffassung der lebenden Natur ist die richtige? und Wie ist diese Auffassung auf die Entomologie anzuwenden?"* Speech in St. Petersburg 1860. Edited by H. Schmitzdorff. St. Petersburg: Verlag der kaiser, Hofbuchhandl, 1864, 237–84.

Battelli, Lorella, Vincent Walsh, Alvaro Pascual-Leone, and Patrick Cavanagh. "The 'When' Parietal Pathway Explored by Lesion Studies." *Current Opinion in Neurobiology* 18, no. 2 (2008): 120–26.

Bauer, Patricia J. *Remembering the Times of Our Lives: Memory in Infancy and Beyond.* Mahwah, NJ: Lawrence Erlbaum Associates, 2007.

Baum, Steve K., Russell L. Boxley, and Marcia Sokolowski. "Time Perception and Psychological Well-Being in the Elderly." *Psychiatric Quarterly* 56, no. 1 (1984): 54–60.

Belot, Michèle, Vincent P. Crawford, and Cecilia Heyes. "Players of Matching Pennies Automatically Imitate Opponents' Gestures Against Strong Incentives." *Proceedings of the National Academy of Sciences of the United States of America* 110, no. 8 (2013): 2763–68.

Bergson, Henri. *An Introduction to Metaphysics: The Creative Mind.* Totowa, NJ: Littlefield, Adams, 1975.

Blewett, A. E. "Abnormal Subjective Time Experience in Depression." *British Journal of Psychiatry* 161 (August 1992): 195–200.

Block, Richard A., and Dan Zakay. "Timing and Remembering the Past, the Present, and the Future." In *Psychology of Time,* edited by Simon Grondin. Bingley, UK: Emerald, 2008, 367–94.

Brand, Matthias, Esther Fujiwara, Elke Kalbe, Hans-Peter Steingass, Josef Kessler, and Hans J. Markowitsch. "Cognitive Estimation and Affective

Judgments in Alcoholic Korsakoff Patients." *Journal of Clinical and Experimental Neuropsychology* 25, no. 3 (2003): 324–34.

Bschor, T., M. Ising, M. Bauer, U. Lewitzka, M. Skerstupeit, B. Müller-Oerlinghausen, and C. Baethge. "Time Experience and Time Judgment in Major Depression, Mania and Healthy Subjects: A Controlled Study of 93 Subjects." *Acta Psychiatrica Scandinavica* 109, no. 3 (2004): 222–29.

Bueti, Domenica, and Vincent Walsh. "The Parietal Cortex and the Representation of Time, Space, Number and Other Magnitudes." *Philosophical Transactions of the Royal Society of London. Series B, Biological Sciences* 364, no. 1525 (2009): 1831–40.

Buhusi, Catalin V., and Warren H. Meck. "Relative Time Sharing: New Findings and an Extension of the Resource Allocation Model of Temporal Processing." *Philosophical Transactions of the Royal Society of London. Series B, Biological Sciences* 364, no. 1525 (2009): 1875–85.

Church, Russell M. "A Tribute to John Gibbon." *Behavioural Processes* 57, no. 2–3 (2002): 261–74.

Church, Russell M., Warren H. Meck, and John Gibbon. "Application of Scalar Timing Theory to Individual Trials." *Journal of Experimental Psychology Animal Behavior Processes* 20, no. 2 (1994): 135–55.

Conway III, Lucian Gideon. "Social Contagion of Time Perception." *Journal of Experimental Social Psychology* 40, no. 1 (2004): 113–20.

Coull, Jennifer T., and A. C. Nobre. "Where and When to Pay Attention: The Neural Systems for Directing Attention to Spatial Locations and to Time Intervals as Revealed by Both PET and fMRI." *Journal of Neuroscience* 18, no. 18 (1998): 7426–35.

Coull, Jennifer T., Franck Vidal, Bruno Nazarian, and Françoise Macar. "Functional Anatomy of the Attentional Modulation of Time Estimation." *Science* (New York) 303, no. 5663 (2004): 1506–8.

Craig, A. D. "Human Feelings: Why Are Some More Aware than Others?" *Trends in Cognitive Sciences* 8, no. 6 (2004): 239–41.

Crystal, Jonathon D. "Animal Behavior: Timing in the Wild." *Current Biology* 16, no. 7 (2006): R252–53. http://www.ncbi.nlm.nih.gov/pubmed/16 581502.

Dennett, Daniel C. "The Self as a Responding—and Responsible—Artifact." *Annals of the New York Academy of Sciences* 1001 (2003): 39–50.

Droit-Volet, Sylvie. "Child and Time." In *Lecture Notes in Computer Science (In-*

cluding Subseries Lecture Notes in Artificial Intelligence and Lecture Notes in Bioinformatics) 6789 LNAI (2011): 151–72.

Droit-Volet, Sylvie, Sophie Brunot, and Paula Niedenthal. "Perception of the Duration of Emotional Events." *Cognition and Emotion* 18, no. 6 (2004): 849–58.

Droit-Volet, Sylvie, Sophie L. Fayolle, and Sandrine Gil. "Emotion and Time Perception: Effects of Film-Induced Mood." *Frontiers in Integrative Neuroscience* 5, August (2011): 1–9.

Droit-Volet, Sylvie, and Sandrine Gil. "The Time-Emotion Paradox." *Philosophical Transactions of the Royal Society of London. Series B, Biological Sciences* 364, no. 1525 (2009): 1943–53.

Droit-Volet, Sylvie, and Warren H. Meck. "How Emotions Colour Our Perception of Time." *Trends in Cognitive Sciences* 11, no. 12 (2007): 504–13.

Droit-Volet, Sylvie, Danilo Ramos, José L. O. Bueno, and Emmanuel Bigand. "Music, Emotion, and Time Perception: The Influence of Subjective Emotional Valence and Arousal?" *Frontiers in Psychology* 4 (July 2013): 1–12.

Effron, Daniel A., Paula M. Niedenthal, Sandrine Gil, and Sylvie Droit-Volet. "Embodied Temporal Perception of Emotion." *Emotion* 6, no. 1 (2006): 1–9.

Fraisse, Paul. "Perception and Estimation of Time." *Annual Review of Psychology* 35 (February 1984): 1–36.

———. *The Psychology of Time.* New York: Harper & Row, 1963.

Fraser, Julius Thomas. *Time and Mind: Interdisciplinary Issues.* Madison, CT: International Universities Press, 1989.

———. *Time, the Familiar Stranger.* Amherst, MA: University of Massachusetts Press, 1987.

Fraser, Julius Thomas, Francis C. Haber, and G. H. Müller. *The Study of Time: Proceedings of the First Conference of the International Society for the Study of Time,* Oberwolfach (Black Forest), West Germany. Berlin: Springer-Verlag, 1972.

Fraser, Julius Thomas, ed. *The Voice of Time. A Cooperative Survey of Man's Views of Time as Expressed by the Sciences and by the Humanities.* New York: George Braziller, 1966.

Friedman, William J., and Steve M. J. Janssen. "Aging and the Speed of Time." *Acta Psychologica* 134, no. 2 (2010): 130–41.

Gallant, Roy, Tara Fedler, and Kim A. Dawson. "Subjective Time Estimation and Age." *Perceptual and Motor Skills* 72 (June 1991): 1275–80.

Gibbon, John. "Scalar Expectancy Theory and Weber's Law in Animal Timing." *Psychological Review* 84, no. 3 (1977): 279–325.

Gibbon, John, and Russell M. Church. "Representation of Time." *Cognition* 37, no. 1–2 (1990): 23–54.

Gibbon, John, Russell M. Church, and Warren H. Meck. "Scalar Timing in Memory." *Annals of the New York Academy of Sciences* 423 (May 1984): 52–77.

Gibbon, John, Chara Malapani, Corby L. Dale, and C. R. Gallistel. "Toward a Neurobiology of Temporal Cognition: Advances and Challenges." *Current Opinion in Neurobiology* 7, no. 2 (1997): 170–84.

Gibson, James J. "Events Are Perceivable but Time Is Not." In *The Study of Time II: Proceedings of the Second Conference of the International Society for the Study of Time, Lake Yamanaka, Japan,* edited by J. T. Fraser and N. Lawrence. New York: Springer-Verlag, 295–301.

Gil, Sandrine, Sylvie Rousset, and Sylvie Droit-Volet. "How Liked and Disliked Foods Affect Time Perception." *Emotion* (Washington, D.C.) 9, no. 4 (2009): 457–63.

Gooddy, William. "Disorders of the Time Sense." In *Handbook of Clinical Neurology.* Vol. 3, edited by P. J. Vinken and G. W. Bruyn. Amsterdam: North Holland Publishing, 1969, 229–50.

———. *Time and the Nervous System.* New York: Praeger, 1988.

Grondin, Simon. "From Physical Time to the First and Second Moments of Psychological Time." *Psychological Bulletin* 127, no. 1 (2001): 22–44.

———. *Psychology of Time.* Bingley, UK: Emerald, 2008.

Gruber, Ronald P., and Richard A. Block. "Effect of Caffeine on Prospective and Retrospective Duration Judgements." *Human Psychopharmacology* 18, no. 15 (2003): 351–59.

Gu, Bon-mi, Mark Laubach, and Warren H. Meck. "Oscillatory Mechanisms Supporting Interval Timing and Working Memory in Prefrontal-Striatal-Hippocampal Circuits." *Neuroscience and Biobehavioral Reviews* 48 (2015): 160–85.

Heidegger, Martin. *The Concept of Time.* Translated by William McNeill. Oxford, UK: B. Blackwell, 1992.

Henderson, Jonathan, T. Andrew Hurly, Melissa Bateson, and Susan D. Healy.

"Timing in Free-Living Rufous Hummingbirds, *Selasphorus Rufusrufus.*" *Current Biology* 16 (March 7, 2006): 512–15.

Hicks, R. E., G. W. Miller, and M. Kinsbourne. "Prospective and Retrospective Judgments of Time as a Function of Amount of Information Processed." *American Journal of Psychology* 89, no. 4 (1976): 719–30.

Hoagland, Hudson. "Some Biochemical Considerations of Time." In *The Voices of Time: A Cooperative Survey of Man's Views of Time as Expressed by the Sciences and by the Humanities,* edited by Julius Thomas Fraser. New York: George Braziller, 1966, 321–22.

———. "The Physiological Control of Judgments of Duration: Evidence for a Chemical Clock." *Journal of General Psychology* 9, (December 1933): 267–87.

Hopfield, J. J., and C. D. Brody. "What Is a Moment? 'Cortical' Sensory Integration over a Brief Interval." *Proceedings of the National Academy of Sciences of the United States of America* 97, no. 25 (2000): 13919–24.

Ivry, Richard B., and John E. Schlerf. "Dedicated and Intrinsic Models of Time Perception." *Trends in Cognitive Sciences* 12, no. 7 (2008): 273–80.

Jacobson, Gilad A., Dan Rokni, and Yosef Yarom. "A Model of the Olivo-Cerebellar System as a Temporal Pattern Generator." *Trends in Neurosciences* 31, no. 12 (2014): 617–19.

Janssen, Steve M. J., William J. Friedman, and Makiko Naka. "Why Does Life Appear to Speed Up as People Get Older?" *Time and Society* 22, no. 2 (2013): 274–90.

Jin, Dezhe Z., Naotaka Fujii, and Ann M. Graybiel. "Neural Representation of Time in Cortico-Basal Ganglia Circuits." *Proceedings of the National Academy of Sciences of the United States of America* 106, no. 45 (2009): 19156–61.

Jones, Luke A., Clare S. Allely, and John H. Wearden. "Click Trains and the Rate of Information Processing: Does 'Speeding Up' Subjective Time Make Other Psychological Processes Run Faster?" *Quarterly Journal of Experimental Psychology* 64, no. 2 (2011): 363–80.

Joubert, Charles E. "Structured Time and Subjective Acceleration of Time." *Perceptual and Motor Skills* 59, no. 1 (1984): 335–36.

———. "Subjective Acceleration of Time: Death Anxiety and Sex Differences." *Perceptual and Motor Skills* 57 (August 1983): 49–50.

———. "Subjective Expectations of the Acceleration of Time with Aging." *Perceptual and Motor Skills* 70 (February 1990): 334.

Lamotte, Mathilde, Marie Izaute, and Sylvie Droit-Volet. "Awareness of Time Distortions and Its Relation with Time Judgment: A Metacognitive Approach." *Consciousness and Cognition* 21, no. 2 (2012): 835–42.

Lejeune, Helga, and John H. Wearden. "Vierordt's 'The Experimental Study of the Time Sense' (1868) and Its Legacy." *European Journal of Cognitive Psychology* 21, no. 6 (2009): 941–60.

Lemlich, Robert. "Subjective Acceleration of Time with Aging." *Perceptual and Motor Skills* 41 (May 1975): 235–38.

Lewis, Penelope A., and R. Chris Miall. "The Precision of Temporal Judgement: Milliseconds, Many Minutes, and Beyond." *Philosophical Transactions of the Royal Society of London. Series B, Biological Sciences* 364, no. 1525 (2009): 1897–1905.

———. "Remembering the Time: A Continuous Clock." *Trends in Cognitive Sciences* 10, no. 9 (2006): 401–6.

Lewis, Penelope A., and Vincent Walsh. "Neuropsychology: Time out of Mind." *Current Biology* 12, no. 1 (2002): 12–14.

Lui, Ming Ann, Trevor B. Penney, and Annett Schirmer. "Emotion Effects on Timing: Attention versus Pacemaker Accounts." *PLoS ONE* 6, no. 7 (2011): e21829.

Lustig, Cindy, Matthew Matell, and Warren H. Meck. "Not 'Just' a Coincidence: Frontal-Striatal Interactions in Working Memory and Interval Timing." *Memory* 13, no. 3–4 (2005): 441–48.

MacDonald, Christopher J., Norbert J. Fortin, Shogo Sakata, and Warren H. Meck. "Retrospective and Prospective Views on the Role of the Hippocampus in Interval Timing and Memory for Elapsed Time." *Timing & Time Perception* 2, no. 1 (2014): 51–61.

Matell, Matthew S., Melissa Bateson, and Warren H. Meck. "Single-Trials Analyses Demonstrate That Increases in Clock Speed Contribute to the Methamphetamine-Induced Horizontal Shifts in Peak-Interval Timing Functions." *Psychopharmacology* 188, no. 2 (2006): 201–12.

Matell, Matthew S., George R. King, and Warren H. Meck. "Differential Modulation of Clock Speed by the Administration of Intermittent versus Continuous Cocaine." *Behavioral Neuroscience* 118, no. 1 (2004): 150–56.

Matell, Matthew S., Warren H. Meck, and Miguel A. L. Nicolelis. "Integration of Behavior and Timing: Anatomically Separate Systems or Distributed Processing?" In *Functional and Neural Mechanisms of Interval Timing*, edited by Warren H. Meck. Boca Raton, FL: CRC Press, 2003, 371–91.

Matthews, William J. "Time Perception: The Surprising Effects of Surprising Stimuli." *Journal of Experimental Psychology: General* 144, no. 1 (2015): 172–97.

Matthews, William J., and Warren H. Meck. "Time Perception: The Bad News and the Good." *Wiley Interdisciplinary Reviews: Cognitive Science* 5, no. 4 (2014): 429–46.

Matthews, William J., Neil Stewart, and John H. Wearden. "Stimulus Intensity and the Perception of Duration." *Journal of Experimental Psychology: Human Perception and Performance* 37, no. 1 (2011): 303–13.

Mauk, Michael D., and Dean V. Buonomano. "The Neural Basis of Temporal Processing." *Annual Review of Neuroscience* 27 (January 2004): 307–40.

McInerney, Peter K. *Time and Experience*. Philadelphia: Temple University Press, 1991.

Meck, Warren H. "Neuroanatomical Localization of an Internal Clock: A Functional Link Between Mesolimbic, Nigrostriatal, and Mesocortical Dopaminergic Systems." *Brain Research* 1109, no. 1 (2006): 93–107.

———. "Neuropsychology of Timing and Time Perception." *Brain and Cognition* 58, no. 1 (2005): 1–8.

Meck, Warren H., and Richard B. Ivry. "Editorial Overview: Time in Perception and Action." *Current Opinion in Behavioral Sciences* 8 (2016): vi–x.

Merchant, Hugo, Deborah L. Harrington, and Warren H. Meck. "Neural Basis of the Perception and Estimation of Time." *Annual Review of Neuroscience* 36 (June 2013): 313–36.

Michon, John A. "Guyau's Idea of Time: A Cognitive View." In *Guyau and the Idea of Time*, edited by John A. Michon, Viviane Pouthas, and Janet L. Jackson. Amsterdam: North-Holland Publishing, 1988, 161–97.

Mitchell, Stephen A. *Relational Concepts in Psychoanalysis: An Integration*. Cambridge: Harvard University Press, 1988.

Naber, Marnix, Maryam Vaziri Pashkam, and Ken Nakayama. "Unintended Imitation Affects Success in a Competitive Game." *Proceedings of the National Academy of Sciences of the United States of America* 110, no. 50 (2012): 20046–50.

Nather, Francisco C., José L. O. Bueno, Emmanuel Bigand, and Sylvie Droit-Volet. "Time Changes with the Embodiment of Another's Body Posture." *PloS One* 6, no. 5 (2011): e19818.

Nather, Francisco Carlos, José L. O. Bueno. "Timing Perception in Paintings and Sculptures of Edgar Degas." *KronoScope* 12, no. 1 (2012): 16–30.

Nather, Francisco Carlos, Paola Alarcon Monteiro Fernandes, and José L. O. Bueno. "Timing Perception Is Affected by Cubist Paintings Representing Human Figures." *Proceedings of the 28th Annual Meeting of the International Society for Psychophysics* 28 (2012): 292–97.

Nelson, Katherine. "Emergence of Autobiographical Memory at Age 4." *Human Development* 35, no. 3 (1992): 172–77.

———. *Narratives from the Crib.* Cambridge, MA: Harvard University Press, 1989.

———. *Young Minds in Social Worlds: Experience, Meaning, and Memory.* Cambridge, MA: Harvard University Press, 2007.

Noulhiane, Marion, Viviane Pouthas, Dominique Hasboun, Michel Baulac, and Séverine Samson. "Role of the Medial Temporal Lobe in Time Estimation in the Range of Minutes." *Neuroreport* 18, no. 10 (2007): 1035–38.

Ogden, Ruth S. "The Effect of Facial Attractiveness on Temporal Perception." *Cognition and Emotion* 27, no. 7 (2013): 1292–1304.

Oprisan, Sorinel A., and Catalin V. Buhusi. "Modeling Pharmacological Clock and Memory Patterns of Interval Timing in a Striatal Beat-Frequency Model with Realistic, Noisy Neurons." *Frontiers in Integrative Neuroscience* 5, no. 52 (September 23, 2011).

Ovsiew, Fred. "The Zeitraffer Phenomenon, Akinetopsia, and the Visual Perception of Speed of Motion: A Case Report." *Neurocase* 4794 (April 2013): 37–41.

Perbal, Séverine, Josette Couillet, Philippe Azouvi, and Viviane Pouthas. "Relationships between Time Estimation, Memory, Attention, and Processing Speed in Patients with Severe Traumatic Brain Injury." *Neuropsychologia* 41, no. 12 (2003): 1599–1610.

Pöppel, Ernst. "Time Perception." In *Handbook of Sensory Physiology.* Vol. 8, *Perception,* edited by R. Held, H. W. Leibowitz, and H. L. Teubner. Berlin: Springer-Verlag, 1978, 713–29.

Pouthas, Viviane, and Séverine Perbal. "Time Perception Depends on Accurate Clock Mechanisms as Well as Unimpaired Attention and Memory Processes." *Acta Neurobiologiae Experimentalis* 64, no. 3 (2004): 367–85.

Rammsayer, T. H. "Neuropharmacological Evidence for Different Timing Mechanisms in Humans." *Quarterly Journal of Experimental Psychology. B, Comparative and Physiological Psychology* 52, no. 3 (1999): 273–86.

Roecklein, Jon E. *The Concept of Time in Psychology: A Resource Book and Annotated Bibliography.* Westport, CT: Greenwood Press, 2000.

Sackett, Aaron M., Tom Meyvis, Leif D. Nelson, Benjamin A. Converse, and Anna L. Sackett. "You're Having Fun When Time Flies: The Hedonic Consequences of Subjective Time Progression." *Psychological Science* 21, no. 1 (2010): 111–17.

Schirmer, Annett. "How Emotions Change Time." *Frontiers in Integrative Neuroscience* 5 (October 5, 2011): 1–6.

Schirmer, Annett, Warren H. Meck, and Trevor B. Penney. "The Socio-Temporal Brain: Connecting People in Time." *Trends in Cognitive Sciences* 20, no. 10 (2016): 760–72.

Schirmer, Annett, Tabitha Ng, Nicolas Escoffier, and Trevor B. Penney. "Emotional Voices Distort Time: Behavioral and Neural Correlates." *Timing & Time Perception* 4, no. 1 (2016): 79–98.

Schuman, Howard, and Willard L. Rogers. "Cohorts, Chronology, and Collective Memory." *Public Opinion Quarterly* 68, no. 2 (2004): 217–54.

Schuman, Howard, and Jacqueline Scott. "Generations and Collective Memories." *American Sociological Review* 54, no. 3 (1989): 359–81.

Suddendorf, Thomas. "Mental Time Travel in Animals?" *Trends in Cognitive Sciences* 7, no. 9 (2003): 391–96.

Suddendorf, Thomas, and Michael C. Corballis. "The Evolution of Foresight: What Is Mental Time Travel, and Is It Unique to Humans?" *Behavioral and Brain Sciences* 30, no. 3 (2007): 299–313; discussion 313–51.

Swanton, Dale N., Cynthia M. Gooch, and Matthew S. Matell. "Averaging of Temporal Memories by Rats." *Journal of Experimental Psychology* 35, no. 3 (2009): 434–39.

Tipples, Jason. "Time Flies When We Read Taboo Words." *Psychonomic Bulletin and Review* 17, no. 4 (2010): 563–68.

Treisman, Michel. "The Information-Processing Model of Timing (Treisman, 1963): Its Sources and Further Development." *Timing & Time Perception* 1, no. 2 (2013): 131–58.

Tuckman, Jacob. "Older Persons' Judgment of the Passage of Time over the Life-Span." *Geriatrics* 20 (February 1965): 136–40.

Walker, James L. "Time Estimation and Total Subjective Time." *Perceptual and Motor Skills* 44, no. 2 (1977): 527–32.

Wallach, Michael A., and Leonard R. Green. "On Age and the Subjective Speed of Time." *Journal of Gerontology* 16, no. 1 (1961): 71–74.

Wearden, John H. "Applying the Scalar Timing Model to Human Time Psychology: Progress and Challenges." In *Time and Mind II: Information Processing Perspectives,* edited by Hede Helfrich. Cambridge, MA: Hogrefe & Huber, 2003, 21–29.

———. " 'Beyond the Fields We Know . . .': Exploring and Developing Scalar Timing Theory." *Behavioural Processes* 45 (April 1999): 3–21.

———. " 'From That Paradise . . .': The Golden Anniversary of Timing." *Timing & Time Perception* 1, no. 2 (2013): 127–30.

———. "Internal Clocks and the Representation of Time." In *Time and Memory: Issues in Philosophy and Psychology,* edited by Christoph Hoerl and Teresa McCormack. Oxford: Clarendon Press, 2001, 37–58.

———. *The Psychology of Time Perception.* London: Palgrave Macmillan, 2016.

———. "Slowing Down an Internal Clock: Implications for Accounts of Performance on Four Timing Tasks." *Quarterly Journal of Experimental Psychology* 61, no. 2 (2008): 263–74.

Wearden, John H., H. Edwards, M. Fakhri, and A. Percival. "Why 'Sounds Are Judged Longer than Lights': Application of a Model of the Internal Clock in Humans." *Quarterly Journal of Experimental Psychology. B, Comparative and Physiological Psychology* 51, no. 2 (1998): 97–120.

Wearden, John H., and Luke A. Jones. "Is the Growth of Subjective Time in Humans a Linear or Nonlinear Function of Real Time?" *Quarterly Journal of Experimental Psychology* 60, no. 9 (2006): 1289–1302.

Wearden, John H., and Helga Lejeune. "Scalar Properties in Human Timing: Conformity and Violations." *Quarterly Journal of Experimental Psychology* 61, no. 4 (2008): 569–87.

Wearden, John H., and Bairbre McShane. "Interval Production as an Analogue of the Peak Procedure: Evidence for Similarity of Human and Animal

Timing Processes." *Quarterly Journal of Experimental Psychology* 40, no. 4 (1988): 363–75.

Wearden, John H., Roger Norton, Simon Martin, and Oliver Montford-Bebb. "Internal Clock Processes and the Filled-Duration Illusion." *Journal of Experimental Psychology. Human Perception and Performance* 33, no. 3 (2007): 716–29.

Wearden, John H., and I. S. Penton-Voak. "Feeling the Heat: Body Temperature and the Rate of Subjective Time, Revisited." *Quarterly Journal of Experimental Psychology. Section B: Comparative and Physiological Psychology* 48, no. 2 (1995): 129–41.

Wearden, John H., J. H. Smith-Spark, Rosanna Cousins, and N. M. J. Edelstyn. "Stimulus Timing by People with Parkinson's Disease." *Brain and Cognition* 67 (2008): 264–79.

Wearden, John H., A. J. Wearden, and P. M. A. Rabbitt. "Age and IQ Effects on Stimulus and Response Timing." *Journal of Experimental Psychology: Human Perception and Performance* 23, no. 4 (1997): 962–79.

Wiener, Martin, Christopher M. Magaro, and Matthew S. Matell. "Accurate Timing but Increased Impulsivity Following Excitotoxic Lesions of the Subthalamic Nucleus." *Neuroscience Letters* 440 (2008): 176–80.

Wittmann, Marc, Olivia Carter, Felix Hasler, B. Rael Cahn, Ulrike Grimberg, Philipp Spring, Daniel Hell, Hans Flohr, and Franz X. Vollenweider. "Effects of Psilocybin on Time Perception and Temporal Control of Behaviour in Humans." *Journal of Psychopharmacology* 21, no. 1 (2007): 50–64.

Wittmann, Marc, and Sandra Lehnhoff. "Age Effects in Perception of Time." *Psychological Reports* 97, no. 3 (2005): 921–35.

Wittmann, Marc, David S. Leland, Jan Churan, and Martin P. Paulus. "Impaired Time Perception and Motor Timing in Stimulant-Dependent Subjects." *Drug and Alcohol Dependence* 90, no. 2–3 (2007): 183–92.

Wittmann, Marc, Alan N. Simmons, Jennifer L. Aron, and Martin P. Paulus. "Accumulation of Neural Activity in the Posterior Insula Encodes the Passage of Time." *Neuropsychologia* 48, no. 10 (2010): 3110–20.

Wittmann, Marc, and Virginie van Wassenhove. "The Experience of Time: Neural Mechanisms and the Interplay of Emotion, Cognition and Embodiment." *Philosophical Transactions of the Royal Society of London. Series B, Biological Sciences* 364, no. 1525 (2009): 1809–13.

Wittmann, Marc, David S. Leland, Jan Churan, and Martin P. Paulus. "Impaired Time Perception and Motor Timing in Stimulant-Dependent Subjects." *Drug and Alcohol Dependence* 90, no. 2–3 (2007): 183–92.

Wittmann, Marc, Tanja Vollmer, Claudia Schweiger, and Wolfgang Hiddemann. "The Relation between the Experience of Time and Psychological Distress in Patients with Hematological Malignancies." *Palliative & and Supportive Care* 4, no. 4 (2006): 357–63.

INDEX

INDEX

Blickfield, 100
Blinking, 116, 126, 137, 211
Blood pressure, 28, 33, 51, 69
Blue-green algae, *see* Cyanobacteria
Body temperature, 27–28, 33, 51–53, 196
Bois-Reymond, Emil du, 110
Boredom, 70, 98, 102, 118, 139, 168, 174, 192–94, 237
Boring, Edward G., 115
Boston Elevated Railway Company, 151
Boston University, 196
Braam, Janet, 32
Brain-imaging studies, 157–60, 213, 219
Brigham and Women's Hospital, 70
British Barbarians, The (Grant), 92–93
Broca, Paul, 111
Brookhaven National Laboratory, 98
Brown University, 202
Buonomano, Dean, 145–50
Bureau International des Poids et Mesures (B.I.P.M.), 4–5, 66
  Consultative Committee for Time and Frequency (C.C.T.F.), 14
  Time Department, 9–15, 235
Business Psychology (Münsterberg), 151
Busyness, 242, 244, 246
Byrd, Admiral Richard, 20, 56

Cabinet magazine, 23
Calendars, 3, 16, 21, 22, 24, 51
California, University of
  Berkeley, 235
  Los Angeles (UCLA), 67, 118, 145
  San Diego, 46, 201, 222
California Institute of Technology (Caltech), 132
Cameo effect, 155–57
Canterbury Tales, The (Chaucer), 189
Carbon atoms, 29
Cardiff University, 233
Carroll, Lewis, 188
Causality, 117, 131, 132, 134–36, 184
Celestial meridian, 103–4, 106
Cerebellum, 222
Cerebral cortex, 176, 224
Cesium, 6–8
Chapman, John Jay, 114
Chaucer, Geoffrey, 189
Chernobyl, 29

Chesterton, G. K., 93
Child's Concept of Time, A (Piaget), 171
Christianity, xii
Chronobiology, 25
Chronometers, 9, 162, 166
Chronoscopes, 102, 105, 112
Church, Russell, 202
Cicero, 5
Cincinnati, University of, 238
Circadian rhythms, 27–42, 67–72, 116–17, 217–18, 232
  animal studies of, 28, 31–34, 38, 45–47, 60, 63
  in Arctic, 56–60, 63–64
  of bodily functions, 27–29, 52–53, 67–69
  fetal, 36–38
  genetics of, 30–31, 32, 34, 36, 37, 46–50, 60, 63, 95
  of infants, 38–43
  in isolation experiments, 24–15, 52–54, 71–72
  microbial, 44–50
  reprogramming of, 35, 70–71
Circular T (monthly B.I.P.M. Time Department report), 12–13
City University of New York, 172
Clay, E. R., 87, 99
Clement, William, 6
Clepsydra (water clock), 5, 202, 221
Climate change, 58, 63, 64
Cognitive science, 102, 117–19, 123
Cold War, 23
Columbia University, 198, 222
Colwell, Chris, 67–69
Concept of Time, The (Heidegger), 249
Condillac, Étienne Bonnot de, 95
Conditioning, 198, 201–2
Confessions, The (St. Augustine), vii, xii–xv
Consciousness, xiv–xv, 16, 80, 111, 131, 134, 142, 166, 220
  Augustine on, xiii, 29–30, 77–80, 86–87, 139, 165, 191, 200, 250
  James on, xv, 78, 83–85, 87–88
  of time, 27, 32, 92, 95–96, 100–103, 106, 119, 123–29, 193, 199, 217, 226
Convention of the Metre, 4
Coordinated Universal Time (U.T.C.), 4–5, 10–14
Corkum, Paul, 97–98

ABOUT THE AUTHOR

Alan Burdick is a staff writer and former senior editor at the *New Yorker*. His writing has also appeared in the *New York Times Magazine*, *Harper's*, *GQ*, *Best American Science and Nature Writing*, and elsewhere. His first book, *Out of Eden: An Odyssey of Ecological Invasion*, was a National Book Award finalist and won the Overseas Press Club Award for environmental reporting.